ECOSYSTEM
MANAGEMENT

Contents

INTRODUCTION

William R. Burch Jr.

People have always managed their critical natural resources with care and respect. However, the definition of what was and was not perceived as a resource *critical* for their survival varies by time, place, and culture. Like other curious, adventurous, and adaptive animals, sometimes members of a given human culture make mistakes and choose the wrong critical resource at the wrong time.

The human culture that experiments and makes the wrong resource choices may bear the consequences of such mistakes in a variety of ways—a forced move to a less damaged ecosystem, becoming dominated by cultures who are more ecologically and/or technologically fortunate, persisting but in great misery, experiencing significant population decline, or having their culture become extinct. Those cultures that do persist develop traditions, stories, and myths that capture the lessons learned and caution future generations about the need to avoid such mistakes.

Though the people of Canada and the United States have made critical resource mistakes—decades of "cut out and get out" forestry in the nineteenth century, and in the twentieth century building on barrier dunes and floodplains, mining groundwater in the Southwest, producing the dust bowl and continuing soil loss in the great plains, converting rich agricultural lands to industrial and residential sprawl—their history seems spared a sense of large-scale ecotragedy. Perhaps the absent historical sense of ecological loss by these rich countries is due to their ability to expand into other people's territory, the low person/land ratio of the continent, their institutional correctives such as environmental legislative action (often more symbolic than functional), or opportune technological adoptions such as coal and petroleum energy converters replacing fuelwood at a most fortuitous time. Whatever the reasons, both countries share a faith that nature and its processes are simply odd problems awaiting the proper rational human control (McPhee, 1989).

In the post–Civil War United States, with industrial capitalism in full optimistic expansion, there was a certainty that the old random, accidental, or traditional responsive learning about natural processes was to be replaced by the steady control of nature through human design and market forces. It was the peculiar mission of the United States to rework from German ideas about forestry the notion that natural resources could and should be managed "scientifically." Nature was to be "disenchanted" and made rational, efficient, and molded into human meaning with the same theories and methods that had created steam engines, factories, and battleships. Nature was a machine (Botkin, 1990) awaiting the appropriate scientific engineering to have it continually provide the maximum output of desired products.

At the start of the twentieth century this vision became the guiding faith of the emerging natural resource bureaucracies, their university-based research and training schools, and their professional associations. National and state forests, wildlife preserves, and parks were increased in number, size, and authority. For nearly seventy years in war and peace, depression and prosperity this American translation of sixteenth century German response to wood scarcity has been the only acceptable approach for research, policy, planning, and management by U.S. natural resource organizations.

This tight, self-confirming faith in scientific management has been joined to military-like organizational structures whose uninformed officials have occupied large segments of the nation's rural hinterland to prescribe, proscribe, regulate, restrict, and manipulate human behavior regarding "their" use of resources. Biophysical science was to provide the means and rationale (and mystique) for maximizing the productivity of outputs as desired by powerful constituency groups. Managers made top-down "scientific" decisions that often overrode local ecological realities and desires and regularly transcended the aesthetics of countervailing constituency groups that saw more in forests and waterways than board feet, or acre-feet of agribusiness irrigation, or potential energy production.

In the late 1950s this cozy pattern was challenged by emerging constituency groups. Their ferment of discontent resulted in a spate of federal environmental legislation in the 1960s and 1970s that fundamentally changed the operational context of natural resource agencies—the Wilderness Act, the National Environmental Policy Act, and the Clean Air and Clean Water Acts. The agencies responded to the changes in their social environment by incorporating new disciplines, new data, and new constituency groups into the old model. Social sciences were simply weaker sisters to some of the softer biological sciences like ecology, yet they still shared a faith in scientific rationality. Indeed, the measures of recreation visitor days, wilderness carry capacity levels, and restrictive user permits all followed the old notion that science could find a way to maximize the production of a limited set of outputs.

Yet as the century crept toward a close, there was an emerging scientific and political concern about the ecological, economic, and cultural sustainability of monocultural resource systems. Further, such systems were seen as a threat to other outputs such as "minor" products, wildlife, and sacred places. Clearly, the logic of implementing practices that maximize biodiversity production will run in a nearly opposite direction to the concern for maximizing board feet or number of deer harvested. Biodiversity compels attention to the entire system, not a few selected products. Biodiversity may actually inhibit traditional economic outputs, and moves toward an ecosystem holism rather than a reductionist approach. So rather than being able to simply co-opt a new constituency into the traditional agency paradigm, the whole way of doing natural resource policy, planning, research, and management was being challenged. With a push from legal actions by Native Americans, even "enchantment" with the earth and its processes began to reemerge as another management alternative. Qualitative measures and human values were given nearly as much respect as allowable cut measures and cost-benefit ratios. Truly, the barbarians were at the gate; some were even already inside.

The chapters in this book document some of the perceptions, strategies, and actions of some natural resource agencies as they seek to respond to the changed reality influencing their policies and practices. Diversity in who participates in resource decisions and in the range of outputs sought is a major change. A second major change is the demand for greater bottom-up participation in the resource decisions. A third major demand is for a more holistic model of how the resource system works and is sustained. Of particular importance in these changes has been an expressed interest for including humans in the ecosystem management model. Though the agencies and their social scientists are seeking positive responses to these changes, their hopes often are larger than their conceptual and actual responses. Many of our chapters document that there remains a certain trained incapacity in natural resource agencies as they seem to accept a rhetorical interest for "including people as a natural component of ecosystems" and then sidestep the full implications in actually trying to make such an inclusion.

Though it is fashionable to talk of the need for new paradigms for the new century, what usually sounds reasonable scans as cliche when it appears in hard print the next day. This book provides ample evidence that the traditional models guiding natural resource agencies and professions have exceeded that point of diminishing returns. The challenges, environments, limits, needs, and hopes of the present and the future are a prism of reality that make the extant theories and tools of delivery seem archaic. Our intent is not to repeat the cliches of the moment but to use the prism of reality as leverage for anticipating and resolving these new challenges. We examine the nature of some of the challenges and responses and identify some tools and ideas from applied environmental social science that may provide more effective and efficient natural resource policy, planning, and management activities.

We do not have a unified guiding theory in this book as organizational ecologists and other social science theorists might offer. Though we respect such attempts, ours is more a collage of empirical snapshots concerning issues and responses affecting a wide array of natural resource agencies and professional activities. These snapshots mostly concern national forestry and park agencies in Canada and the United States. As such, the snapshots are not a complete enough sample for full domestic or world theory testing. We would be imprudent and more than a little arrogant to assume that we could hope to follow leaders such as Hannan and Freeman (1989, xi), who have "an approach to the macrosociology of organizations that builds on general ecological and evolutionary models of change in populations and communities or organizations. The goal of this perspective is to understand the forces that shape the structures of organizations over long time spans."

One reason that we do not follow the organizational ecology model of Hannan and Freeman and their present-day followers is that, like Perrow (1986, 213), we think that the model "neglects the fact that our world is made in large part by particular men and women with particular interests. Instead, it searches for ecological laws that transcend the hubbub that sociology should attend to." Ours are reports from the front lines. There is more than a lot of sensitivity to the hubbub of social life as it is actually practiced. Many of our authors are the women and men making the visions, hopes, and solutions to the natural resource challenges facing all of the agencies. We do not study the problem as some abstract entity that advances some arcane theory (no matter how important this may ultimately be) but rather as persons seeking to understand, to resolve, and to ensure a more sustainable and harmonious relationship between humanity and the nonhuman world. For the most part, then, we are natural resource professionals first and social scientists more out of the necessity to make things work than to solve a particular academic theorem.

We do use the term "adaptive," meaning to adjust to changing realities. However, it is not in some technical Darwinian sense for explaining the behavior of a given species or organization. We mean that natural resource professionals and agencies seek ways to adjust to their emerging challenges. They do this with new visions such as "participatory management," "rapid rural appraisal," "new perspectives in forestry," and "ecosystem management." Our authors examine some of these processes to help improve the policy, planning, and management of ecosystems, not to analyze agencies or their professionals as part of an ecological metaphor. So this book uses some social science to help natural resource managers do a better job and to adapt to or meet the new challenges, but it is not a full-fledged social science tome. Those analyses we leave to our sisters and brothers in the traditional academic departments.

Indeed, our authors might have some unease with elegant social scientific metaphors because many have been busy experiencing the whiff of a newly started chain saw, parting the wolf willows along the riparian zone of a meadow to check grazing influences on fisheries, interviewing campers as to their hopes and needs, working with an inner city neighborhood in developing a community forestry project, paddling a canoe in high wind with a group of urban scouts, planning a nature center, riding with cowboys through public range sagebrush, and tracking poachers of endangered species. Those readers who like their social science pure and unadulterated are likely to find this volume too pedestrian; on the other hand, those who like considerable helpings of application with their social science might find this a pathway very much to their liking.

Regardless of your particular scientific preference, we hope you will take our efforts as a start on confronting some of the necessary natural resource and environmental questions of the coming decades. If ecosystem management is to become one of the means for dealing with those critical questions, it will require a great deal more serious attention than it has yet received. Ours is a start on that exploration.

REFERENCES

Botkin, Daniel B. 1990. *Discordant harmonies—A new ecology for the twenty-first century.* New York: Oxford University Press.

Hannan, Michael T., and John Freeman. 1989. *Organizational ecology.* Cambridge, Massachusetts: Harvard University Press.

McPhee, John. 1989. *The control of nature.* New York: The Noonday Press.

Perrow, Charles. 1986. *Complex organizations—A critical essay.* 3d ed. New York: McGraw-Hill.

Part I

Emerging Challenges for Natural Resource Organizations in the Twenty-First Century: Problems, Politics, and People

Overview

Natural resource professionals might assume that after the 1960s and the 1990s they have seen all of the problems from all of the political directions all of the time. After all, 1960s legislative action greatly restricted the latitude of decisions by professionals but left them to bring the bad news to the local people. And then, just as they adapted to these massive changes, the 1990s try to reverse the 1960s by taking back all the money and personnel, which starves the natural resource agencies of their necessary resources for meeting the legislative demands of the 1960s. Yet, in the decades to come these times are likely to be points of nostalgic remembrance compared to the issues now emerging.

Our authors cover four of these coming challenges and suggest some possible responses. These challenges range from overlooking the possibility that changes in the global economy will be the predominant influence upon plans for local community stability, to changes in dominance hierarchies where metro centers can no longer impose their will upon the rural hinterland, to parks as major hard currency earners (Yen, Marks, Krone) and the need to adapt to foreign tourist needs, to changing demographics of forest visitors, to emerging stakeholder groups whose desires and needs have seldom been addressed by resource agencies. Of course, one could identify many other challenges, but these seem the most critical in terms of the need for major changes in the agencies.

Our authors challenge the agencies to question their preparedness for these critical changes. How have they incorporated response to global change in their plans and strategies? How prepared are they to respond to localities now occupied with well-educated populations who know how to work the system? How prepared are they for serving as destination points for foreign tourists from much richer situations than those of the locals? How prepared are they for the switching from serving the youth of wilderness testing to the elderly with a more comfortable approach to wilderness? How prepared are they to meet the needs of environmental justice? Certainly, when we hear that "humans are part of the ecosystem," we must wonder if these questions will therefore become part of ecosystem management in the coming decades.

This section should amply demonstrate some of the major challenges to the old ways of doing business by natural resource professionals and agencies. Of particular interest is the fact that most of the ''perturbations'' to the ecosystem are shaped by the contours of change outside the usual domain of resource agencies; yet only small attention to these changes has been paid by the agency research, policy, training, and management programs that should be anticipating and dealing with the emerging challenges. This book considers some of the constraints in vision and some of the possible responses in tools, techniques, and organizational change.

William R. Burch Jr.

Chapter 1

New Forestry, Neopolitics, and Voodoo Economies: Research Needs for Biodiversity Management

Gary E. Machlis

The 1990s are producing a series of important changes in American thinking about natural resource management. Some of these changes reflect deeply significant shifts in paradigms (new ways of thinking). Others are superficial rhetorical changes (new ways of speaking). While it is important to separate substance from style, both influence the research needs of agencies charged with biodiversity conservation, and both influence what the scientific community can and will deliver.

For the conservation of biological diversity, three paradigm shifts are important: a new forestry, new politics, and new economies. Yet the new forestry may be old, the politics illusory (neopolitics), and the economies based on faith (voodoo economies). It is because of these paradigmatic and rhetorical shifts that managers find themselves searching for new research and applied results, what Lindblom and Cohen (1979) call "usable knowledge." The purpose of this discussion paper is to evaluate current and needed usable knowledge related to these three major shifts in thinking and to use biodiversity conservation as an example of trained incapacities within the natural resource field. The focus is on the human dimensions of biodiversity management and the potential contribution of the social sciences to a biodiversity research agenda.

DEFINITIONS

Because biodiversity is a biological variable and the driving forces leading to biodiversity loss are social, political, and economic, the challenge of providing relevant research findings is clearly interdisciplinary. Cross-disciplinary communication is often confused by unclear concepts and ill-defined terms. Hence a brief set of definitions for biodiversity and the social sciences may be useful.

Biodiversity

There are numerous definitions of biological diversity; most treat diversity as a qualitative state at the genetic, species, or ecosystem levels. Within population biology and ecology, diversity is a multidimensional concept having three components: (1) the number of species coexisting within a uniform habitat ($\propto$-diversity), (2) species turnover rates (β-diversity), and (3) species turnover rate within similar habitats in different regions ($\bullet$-diversity). Wilcox (1984) defines biodiversity as "the variety of life forms, the ecological roles they perform, and the genetic diversity they contain." A definition by the U.S. Office of Technology Assessment (OTA) is similar:

> Biological diversity refers to the variety and variability among living organisms and the ecological complexes in which they occur. Diversity can be defined as the number of different items and their relative frequency. For biological diversity, these items range from complete ecosystems to the chemical structures that are the molecular basis of heredity. Thus, the term encompasses different ecosystems, species, and genes, and their relative abundance (1987, 1).

There is disagreement over how to measure biological diversity; current measures select different components of ecosystems for emphasis. Potential indicators include number of species or "richness," abundance and distribution of populations, number of endangered species, centers of species richness with high endemism, taxic diversity, and degree of genetic variability. Other approaches treat ecosystem functions, interactions, natural communities, successional stages, or ecological redundancy as key measures of diversity.

There is widespread agreement that global biodiversity is being reduced at an accelerated rate; there is less agreement about the actual level of biodiversity loss, compounded by the wide range of operational measures, variation between biomes, and lack of baseline knowledge concerning the number of species and taxa. Even when there is a common measure (say, species richness) and biome (tropical forests), large differences exist in estimated rates. Estimates of species extinction in the tropics range from 15 to 50 percent of all species by the year 2000.

Using any of these estimates, the loss of biodiversity is an important social and biological concern. Lost species may have commodity values; an example is the rosy periwinkle, a single species of plant yielding pharmaceuticals worth

$100 million annually. Species may have intrinsic values within a sociocultural system, and landscapes that provide habitat may have symbolic meanings, provide an economic resource base, or serve legally mandated functions such as wilderness. Biologically, species loss may lead to synergistic effects upon other species, altered energy flows and nutrient cycles, reduced "ecosystem services" such as oxygen production and climate modification, and a decline in ecosystem resilience. Low-diversity ecosystems (such as European heath lands) may make important contributions to local biodiversity, and "hyperdiverse" ecosystems (such as the Amazonian and Asian tropical forests) play a critical role in overall global biodiversity.

The Social Sciences Described

A history and description of the social sciences is neither possible nor necessary here, but a primer on the scope of the social sciences may be useful. Orthodox approaches place six disciplines in the social sciences: anthropology, economics, geography (human rather than physical), psychology, political science, and sociology. History is marginally excluded. Contrary to conventional wisdom, the social sciences are not particularly young: economics, for example, long precedes the development of modern chemistry, and most of the social sciences precede ecology.

These sciences have much in common: research techniques such as observation, social surveys, and experiments are used in all. Boundaries between the sciences are nebulous and prone to arcane distinctions; subfields such as social psychology and economic sociology flourish in academe. New specializations emerge yearly, tracking the growth of knowledge (some of it usable knowledge) and the search for "relevance," funding, or both.

For those interested in understanding the human dimensions of biodiversity conservation, what may be useful is a comparison of each discipline's special focus, i.e., where the discipline has traditionally concentrated intellect and effort. A "map" of the social sciences can be described in general terms. Table 1 provides a basic outline, organizing the sciences around their key units of analysis (the scale of things they study) and the central "engine" of change (the driving forces considered most important).

Anthropology focuses primarily upon social groupings that are intensely cultural: communities, subcultural groups, and even entire cultures themselves. The driving forces are primarily cultural change, with the role of tradition being a critical interest. *Economics* (which could be split into macroeconomics and microeconomics) treats markets, industries, and economies as key units of study; the driving force of change is economic value (broadly defined). *Geography,* (specifically, human geography) treats regions, landscapes, and other spatial units (governmental, environmental, and so forth) as critical, and the spatial distribution of people, resources, and culture is seen as a significant driving force.

Table 1   Basic Outline of the Social Sciences

Discipline	Key units of analysis	Engines of change (driving forces)
Anthropology	Communities Subcultures Cultures	Tradition and culture
Economics	Markets Industries	Economic value
Geography	Regions Landscapes	Spatial distribution
Psychology	Individuals	Communication
Political science	Institutions States	Power
Sociology	Social groups Organizations Communities	Conflict and cohesion

Psychology's key unit is the individual, and communication of meaning (within and between individuals) is a central driving force. *Political science* focuses upon the institutions of state (at many levels); the central engine of change to many political scientists is power and its use. *Sociology* treats social groups, organizations, and communities as key units of analysis, with conflict (social disorder) and cohesion (social order) as central forces driving change.

Several patterns emerge. First, the social sciences overlap considerably as to their units of analysis; no clear demarcation exists between, for example, anthropology and sociology applied at the community level. Second, the social sciences have remained largely unfamiliar and uncomfortable with ecosystem- and global-level analyses. Third, these disciplines reflect the complexity of human social behavior: tradition, value, power, and space are critical to understanding the human condition. If the social sciences can help to accurately describe the mechanisms that link social and biological systems, and help in predicting these interactions, they can make a significant contribution to biodiversity research and management.

THE THREE SHIFTS OF PARADIGM AND RHETORIC

The three shifts referred to earlier—new forestry, new politics, and new economies—are directly important for setting an applied research agenda related to biodiversity, for they are changing the amount and kind of scientific information needed by natural resource managers.

Each of the shifts has several components. There is the *paradigm shift* itself, i.e., a significant change in the assumptions, perspective, or world view of managers. The paradigm shift is associated with a *new rhetoric,* as terms and concepts are juggled to express (or confuse) the changes in thinking. For each shift there is a set of *key institutions* (communities, agencies, interest groups, and so forth) that respond most directly to the changing set of accepted ideas about resource management. Each shift is driven partly by a *biodiversity "grail,"* or an objective idealized and sought after by managers. To approach this grail, managers need answers to specific *applied research questions,* and each of the three shifts is associated with a critical question. Finally, each critical research question is currently (or could be) addressed by both a basic scientific discipline and several *research areas.*

New Forestry

New forestry emerged from Jerry Franklin's research on mature temperate forests in the Pacific Northwest, from Likens and Bormann's Hubbard Brook studies of material flows in a northeast watershed, and from a revival of European silvicultural techniques aimed at sustained yield and mixed age-class forest stands. The shift in thinking can be stated simply: new forestry places the dominant emphasis on the ecological system left in place after harvest, as opposed to the economic outputs derived from harvest. The paradigm shift is a significant departure from the management principles that lead to clearcutting strategies, which in turn, have ravaged parts of the western landscape (including severe damage to riparian zones). Yet many of the forest management techniques appropriate to the new way of thinking (such as selective thinning and buffer corridors) have been traditionally taught to professional foresters since the 1950s. New forestry is partly old forestry expressed in new terms and with new importance.

The shift in rhetoric is important for its indication that scale issues are widely viewed as critical. The most significant change in language is the concept of "ecosystem management." Ecosystem management has become de rigueur in environmentalist parlors and industry headquarters, and resource management agencies from the National Park Service to the Environmental Protection Agency have claimed it as an appropriate strategy. Yet the term is so ill-defined as to be misleading, and the difficulties of actually practicing ecosystem management are extreme.

There are several reasons for caution: (1) ecosystem boundaries are difficult to describe, (2) ecosystems can vary in size and scale from microsites to the planet, (3) variation in ecosystem types is graduated rather than distinct, and (4) knowledge of ecosystem functioning is more theoretical than empirical. The rhetorical shift does show concern for managing at scales larger than forest stands or watersheds, and social experiments such as the Greater Yellowstone Ecosystem illustrate the political and managerial appeal of the term.

If ecosystem management is to be practiced in situ, the key institutions are likely to be resource management agencies, governments (county, state, and federal), and corporations with large land holdings. For these institutions the objective related to biodiversity is likely to be a search for the "best management practices" (BMPs) for biodiversity. BMPs will be those that maximize the conservation of biological diversity while accomplishing other management objectives (timber production, recreational opportunities, fish landings, and so forth). Hence, the critical applied research question is, Which practices are best?

This question can best be served by several applied research areas. Ecosystem evaluation and environmental risk assessment will be crucial, for these approaches can accommodate comparative studies of alternative management practices within a particular ecosystem type. Experience is high in risk assessment related to industrial practices (such as pollution releases and single sites); it is less developed for the study of renewable resource practices and activities at ecosystem and landscape scales.

The foundation for such research is systems ecology, and theoretical advances in this field related to ecosystem stability and resilience have been slow. Systems ecology has not dealt extensively with biodiversity at scales larger than communities; Figure 1 shows four alternative hypotheses regarding the relationship between diversity and ecosystem functioning. For example, curve A suggests redundancy in species function, and curve C suggests a curvilinear function whereby further increases in diversity may actually reduce ecosystem function. Evidence has not accumulated to contradict any of the alternatives; the need to do so as a basis of applied biodiversity research is obvious.

The question of best management practice is socioeconomic as well as ecological. Research areas such as economic contingency value evaluation and social impact assessment will be crucial. Contingency value estimation techniques are crude and imprecise ("How much are you willing to pay to preserve the spotted owl?"), and their weaknesses as compared to classical cost-benefit analysis are significant. The technique's value lies in including noneconomic values, a crucial component of biodiversity. Hence, it will continue to be employed and (hopefully) improved. Social impact assessment is well advanced with a large set of applications, yet its broad methodologies and often time-consuming fieldwork make its utility in assessing BMPs less than ideal.

An important prospect for this major shift in thinking may be to develop a synthesis of environmental risk assessment, economic analysis, and social impact assessment. Managers could apply this holistic approach to alternative management practices within ecosystems where biodiversity conservation is an important objective. Decision makers could select BMPs with more scientific information at hand.

Systems ecology is, of course, a critical discipline for research in this area. From the map of the social sciences described earlier, it is clear that geography, economics, and sociology are well poised to make specific contributions toward answering the critical research questions.

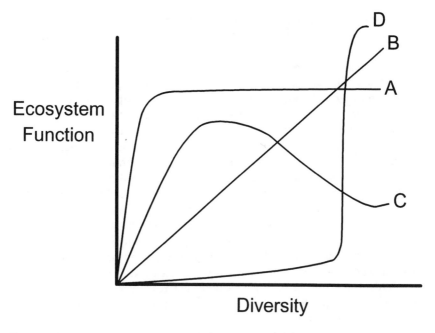

Figure 1.   Alternative relationships between ecosystem function and diversity.

New Politics

The postwar national politics of the United States bears little resemblance to the efficient and polite democracy of our grade-school textbooks, the Boy Scout handbook, or the nostalgic (and forgetful) remembrances of our grandparents. There has been an explosive growth of special interest influence: in 1961 there were 365 registered lobbyists in the District of Columbia; in 1987 there were 23,011. Environmental groups are a good example: triggered by opposition to Jim Watt and others, the Wilderness Society grew from 50,000 members in 1981 to 190,000 members in 1988. Congressional staffers exert technical and political influence. Scandals such as the savings and loans crisis simply expose the commonplace as the nature of American politics in the 1980s turned from the civic to the selfish.

Ironically, the contemporary democratic regime created new public involvement processes coincident with the growth of special interest influence. The new rhetoric of ''public involvement,'' ''public participation,'' and involving ''stakeholders'' represents, at its most cynical, an effort by special interests and corporations to co-opt citizens and interpenetrate government agencies. This modern form of American democracy, where no one is elected and citizens do not vote, is neopolitics, i.e., a new and derivative form of democratic regimes. Environmental issues such as biodiversity management are not immune from its effects.

The key institutions involved in the new public participation are, oddly enough, governments and special interest groups. The National Environmental

Policy Act was crucial legislation, for it required public involvement, and Congress has, for example, increased the level of required public participation in every modification of the Endangered Species Act (and its forerunners). Government agencies such as the U.S. Forest Service and Bureau of Land Management have created policies and processes to integrate or co-opt public views into their decision making. It is now commonplace in the newspapers of western towns to find "invitations" to "public involvement hearings" from various agencies, and sometimes government employees outnumber the citizenry. Special interest groups from preservation groups to industrial associations have become adept at using public involvement processes to further their strategic objectives, gain awareness and membership, and prepare the grounds for legal challenges.

To resource managers charged with biodiversity conservation, the objectives or biodiversity grail of neopolitics are to (1) satisfy agency requirements, (2) garner actual participation from a wide set of stakeholders, and (3) provide the basis for "suit-free" decisions. That is, neopolitics is most successful when it wards off the legal process. The neopolitics of resource management is seen by many managers as both dangerous (their technical skills may be insufficient) and burdensome (the activities take time and resources). Hence, the critical applied research question is, How can we efficiently and democratically make resource management decisions?

This question can best be answered by research in several applied areas. First, research on conflict resolution and mediation has potential to provide alternative processes that can "avoid train wrecks" (Secretary Babbitt's restatement of this biodiversity grail). Unfortunately, the scientific research in this area is sparse. Mainstream sociology has developed a large body of theory and evidence regarding social conflict and its resolution (going back to Sherif's studies of intergroup processes in the 1950s), but there has been relatively little effort to apply broader knowledge of conflict to the issues of resource decision making related to biodiversity.

Second, there is a growing literature on policy formation (primarily from political science but also from economics), and while the majority of the work has dealt with social policies such as education, welfare, and health care, some has effectively been focused on resource management and environmental issues. Tobin's work on the evolution of the Endangered Species Act is illustrative and shows the agencies to be surprisingly active in the early stages of policy formulation, then turning resistant and ineffective in later stages of enacting policy change. Further research is needed, primarily in the arena of understanding the particular role locals play in federal land policy, i.e., the principle of "home rule."

Third, research on public attitudes and values has importance to those managing symbolic landscapes (such as Yellowstone) and "charismatic species" (such as the wolf and grizzly bear). Kellert's work in the 1970s, which involved the systematic cataloging of public attitudes toward individual species, provided the first contemporary hint that species have divergent meanings. The public is divided in their beliefs toward wildlife and the charismatic attraction of certain animals that can trigger public action. Recent exposés by Bonner (on the African elephant) and Schaller (on the giant panda) have shown the cynical use of these insights by interest groups, primarily for raising funds. The inventorying of opinions needs to be augmented by studies of factors influencing public perception, particularly the role of the media in public attitudes toward land management. The spotted-owl/old growth struggle in the Pacific Northwest is an excellent case study waiting to be undertaken. Environmental historians, if involved at the early stages of research, would likely provide valuable insights.

To a lesser degree, research areas such as environmental education and legal studies could provide assistance to managers attempting to make efficient and democratically inspired decisions. Evidence regarding public education processes has been well advanced on single, relatively simple issues such as recycling behavior, but work directly related to complex issues such as biodiversity is sparse to nonexistent. Legal studies (not case law) hold promise if they can be closely linked to the policy research described earlier and help managers better understand the extralegal processes that influence the legal system.

In better understanding the new politics, political science, anthropology, and psychology may offer particularly crucial contributions.

New Economies

A major shift in paradigm and rhetoric has occurred in the objectives of economic planning related to natural resource management (particularly forest management). From WWII to the 1960s, economic development was defined as growth in gross national products (GNPs); the necessary conditions and stages of economic growth were considered to be standard across nations. Forest management served development to the extent that it produced inputs into a GNP. Alternative approaches began to emerge in the 1960s, as the failures of the earlier strategies became undeniable. Economic development was redefined as an increase in per capita improvements in living conditions (employment, health care, education, and so forth). Forestry objectives shifted to a mix of outputs: maximizing returns on capital (of course), but in addition, providing jobs and preserving wildlife, water quality, soil, and so on. This is a real and significant shift in the forester's paradigm.

Simultaneously, there was a fashion cycle of forestry and development rhetoric, both internationally and domestically. In the late 1970s, "social forestry" emerged as a management concept, the term imported from India. Social forestry metamorphized in the United States to become "community forestry," briefly became oriented around "community stability," and was repackaged (coincident with the 1992 Earth Summit) as "sustainable development" and "sustainable forestry." These shifts are largely rhetorical. A consistent and coherent definition of sustainable development has not emerged, and the questions, *What is* sustained, for *whom* is it sustained, and for *how long* is it sustained?, are often avoided.

Managers' concerns reflect the desire for new economies, as opposed to simply a new economic strategy. Local, rural communities have become key units of concern and, in actuality, reflect small towns in mass economies, linked to regional, national, and global economic systems by the flow of resources, labor, and capital. In addition, the new paradigm and rhetoric place increasing concern on "industries" that are more like fragmented economic sectors—the "tourism industry" is an example. Marginalized firms (such as value-adding specialty mills) are important to these economies, and they treat occupational groups (such as loggers) as critical components in need of preservation.

With sustainable development through new rural economies as a desired outcome, it is clear that one biodiversity grail of most resource managers is some form of community stability and new opportunities for creating profit and jobs. Stability may be defined by civic boosters as continued economic expansion, and therefore the pressure is on resource managers to find new ways that forest ecosystems can produce economic value. Hence, the critical applied research question is, How can biodiversity management create sustainable economies?

There are several applied research areas that can support the quest for answers and guidance. Surprisingly, classical economics has limited value, for traditional approaches to cost-benefit reflect a willingness to externalize environmental and social costs and treat the movement of capital as an efficiency regardless of local impact. The above management questions require a form of economic analysis that widens cost-benefit factors to include noneconomic factors and treats movement of capital and jobs as a negative outcome. Research on economic diversification could be useful, and while there is significant experience related to eastern U.S. cities and industrial plant closings, there is little empirical evidence on diversifying small resource-dependent communities in the West. Much of the discussion on diversification revolves around the economic-generating capacity of wilderness, recreation areas, and tourism; the lack of evidence and extent of confidence suggest "voodoo economics" at work. Such research will require numerous case studies for patterns to emerge.

Another area of research with promise is the sociological analysis of resource dependency. This work, which began in the mid-1980s, reflects the practical need to understand the relationship between resource production and community social change; instead of net cost-benefit, the focus shifts to who benefits and who loses. The need to better link economic and sociological analysis is obvious; economists and sociologists have simply not cooperated on research.

Other research areas could benefit managers and serve to "de-voodoo" sustainable development. The research and development (R&D) of new technology and value-adding processes (appropriate to rural scales) could provide a stream of economic alternatives. Such R&D must include analyses of job-creating potential, long-term supply requirements, and environmental impacts. Social

Table 2. "Map" of the issues/research needs related to biodiversity management

Paradigm shift	New rhetoric	Key institutions	Biodiversity "grail"	Applied research question	Research areas
New forestry	Ecosystem management	Resource management agencies, industries	Best management practices	Which practices are best?	Ecosystem evaluation Risk assessment Contingency evaluation Cost-benefit analysis Social impact assessment
New politics	Public participation	Governments and interest groups	Wide public involvement and "suit-free" decisions	How can we efficiently and democratically make decisions?	Conflict mediation and resolution Public environmental education Legal studies Public policy formation Public attitudes/ values
New economies	Sustainable development	Local communities and regional economies	Community stability and new forms of profit taking and jobs	How can biodiversity management create sustainable economies?	Rural/subsistence economies Economic diversification Community stability Technology R&D Social impact assessment

impact assessment (discussed above) could be useful in understanding the consequences of economic change. Analyses of subsistence uses, shadow economies, and black markets would capture the reality of small-town economies and their impact on wildland ecosystems.

Economics, sociology, and geography can make particularly important contributions toward understanding the new economies and their relationship to biodiversity conservation.

Conclusion

American resource management is in transition, and three shifts are critical: new forestry, new politics, and new economies. Table 2 summarizes the previous discussion and illustrates the range of research areas that can support biodiversity conservation and management.

Even if a coherent and sustained biodiversity research program is begun and eventually completed, there are significant barriers to its adoption, diffusion, and use. Contrary to their expectations, scientists' work is not necessarily seen by resource managers as an asset. Important and often overlooked, a structural tension exists between science and resource management.

When the scientist presents research results, the implications lead to a narrowing of managerial options. If the scientist goes so far as to make recommendations, this narrowing is even more pronounced, for to ignore the scientific recommendation is to become vulnerable to lawsuit, reversal of decision, and public controversy—all anathemic to most practicing resource managers. The strategic manager wants a wide range of options; the responsible applied research scientist wants to identify the most effective options and thereby eliminate some resource practices as unwise.

Reluctance on the part of the manager, annoyance on the part of the scientist, and tension between them is always likely and perhaps inevitable. Science and natural resource management each have their own ''culture'' and self-interests. Their interaction is part substance, part style, both paradigm shift and rhetorical fashion. The reality of biodiversity loss, and its menacing potential for ecological harm, grows regardless.

REFERENCES

Lindblom, C. E., and D. K. Cohen. 1979. *Usable knowledge: Social science and social problem solving.* New Haven, CT: Yale University Press.

United States Office of Technology Assessment. 1987. *Technologies to maintain biological diversity.* Washington, DC: Government Printing Office.

Wilcox, B.A. 1984. Concepts in conservation biology: Applications to the management of biodiversity. In *Natural diversity in forest ecosystems,* edited by J. L. Cooley and J. H. Cooley. University of Georgia, Institute of Ecology, Athens, Georgia, p. 640.

Chapter 2

International Tourism: Current Trends and Market Research with Implications for Managing Public Attractions

Joseph T. O'Leary

Scott M. Meis

Changes are occurring in North American travel that will influence economic conditions, destination choices, and travel product offerings. Natural resource agencies play a major role in the opportunities that are being presented. However, recognition of new and emerging customer or visitor groups, developing partnerships, and improving communication with tourism groups are all challenges faced by resource organizations in the future.

Six years ago, a five-year tourism research agreement was signed between Canada and the United States. This past year that agreement was renewed for another five years. A key clause of those agreements was a commitment of the tourism agencies of both nations to work together to understand international tourism within the two countries. A second key clause was a commitment to exchange information on the significance of natural and cultural resources relevant to international tourism in the two countries.

Since the beginning of these joint research agreements, we have jointly studied the international travel markets to North America from some fourteen nations. In 1989, studies of the big four overseas markets—France, Germany, United Kingdom, and Japan—were replicated.

There have also been independent studies of the domestic and international travel markets within and between our two countries. Canada carried out two such major studies of the U.S. pleasure travel market, one in 1985 and a second in 1989. The United States, in turn, has in the past year conducted a study of the Canadian outbound travel market in which Tourism Canada participated as a cosponsoring partner along with Las Vegas, Nevada, and Mexico.

At the time that this cooperative international research program was initiated, cooperation also began between the Canadian Parks Service and Tourism Canada. In recognition of the information potential of the international tourism research program for the Canadian and U.S. national park services, two parallel agreements were established, one between the South east region of the National Park Service and the Canadian Park Service and a second between the University of Waterloo and Clemson University. Through these parallel agreements, influence was exerted on the objectives and content of the joint tourism research program between the two countries to the potential benefit of land and cultural resource management agencies in both countries. Much of the material we present has been derived from the results of these original cooperative structures and the personal and professional exchanges and collaborations that have evolved from them.

The key objectives of this chapter are to provide the following:

- new insights into the evolving trends in tourism relevant to North America
- an understanding of the range of new studies and information on the subject of tourism and its relation to natural and cultural areas
- an appreciation of the size, significance, and characteristics of natural and cultural market segments within international tourism markets in North America
- an appreciation for the implications of this tourism marketing research for managing public resource attractions in North America

DEFINITION OF TOURISM

The World Tourism Organization states that a tourist is

> any person visiting a country (or region) other than that in which he has his usual place of residence . . . (for purposes of personal pleasure)

There are a number of other ways of defining the scope of tourism, some dealing with the activities people engage in, others by the distance or length of time away from home, and others still by the goods and services people make use of in the course of their travels. For our purposes at the moment, however, this definition is sufficient.

In Canada, criteria are normally operationally defined in terms of pleasure travel greater than eighty kilometers, or approximately sixty miles, and involving more than a single day's excursion. In the United States, on the other hand, the normal distance criterion is one hundred miles, although there are some organizations beginning to change this definition to something that approximates the Canadian definition. Pleasure travel involving smaller durations or distances is generally treated as day excursion pleasure travel, not tourism. Culture and/ or nature or outdoors tourism then is defined as any of this sort of long-distance pleasure travel that involves the use of natural or cultural resource amenities.

INTERNATIONAL MARKET TRENDS

Tourism is a major economic factor throughout North America. In 1990, tourism ranked fourth among Canadian exports (arrival of international visitors treated as an export) and was responsible for $26 billion in total receipts, $7.4 billion in foreign exchange earnings, $16.8 billion in direct income, and $11.3 billion in government revenues. In the United States, international visitation has grown about 55 percent in the last five years, going from about twenty-six million visitors in 1985 to an estimated forty million in 1990. Receipts from travel to the United States also grew significantly during this period, going from $22 billion in 1985 to almost $53 billion in 1990. In the United States these numbers have caused quite a stir, since they represent a positive balance of trade associated with tourism.

But travel is also changing. In both Canada and the United States, domestic travel appears to have peaked in about 1988 and then leveled off. U.S. tourists to Canada appeared to have reached a peak in 1986 and then declined through 1990. At the same time this was occurring, the number of Canadian tourists to the United States showed a steady growth, while Canadian visits overseas also showed a sharp increase. For Canadians, many trips were to border states with some shift away from travel to states farther away. To counterbalance some of these changes, overseas travel to Canada began to sharply increase each year beginning in 1986. This followed the same trend noted earlier for the United States.

In the early 1990s, market growth slowed. This seems to be related to the economy and depressed consumer confidence on the domestic North American side, although outbound travel from North America has continued to grow. Economic changes throughout the world have also caused a slowdown in travel from the major international markets like Germany and Japan.

The leading edge of travel in the world appears to be in Asia, certainly in terms of interregional activity. However, new destination areas have also emerged in the region that are providing competition for North American destinations. The United States also appears to be the first choice of Asian travelers when they look outside their own region for a place to visit.

If we consider the economic "news" reported earlier about the importance of tourism in the economy of the two countries, changes that influence people to either not travel or do it differently can have profound impacts on regions, provinces, states, and local communities. Although some resource managers might like to see travelers stay at home, the resources they care for still represent major attractions for tourists throughout North America and jobs for people in the local area. If travel behavior changes, these relationships are thrown out of balance. To address these shifts, understanding travel behavior becomes more critical than ever for resource organizations as well as for other groups.

RECENT MARKET STUDIES OF SIGNIFICANCE

There have been three large research programs within the last several years designed to improve our understanding of the behavior of Canadian, U.S., and international pleasure travel markets. Beginning in 1986, fourteen studies have been conducted examining long-haul pleasure travel outside the country of residence. This research has included countries in Europe, North America, Asia, and South America. Personal interview questionnaires were collected in each country and gathered information on sociodemographic characteristics, travel characteristics, activities engaged in, information sources used to plan a trip, travel philosophy and benefits pursued, places visited on most recent trip, and feelings toward U.S. and Canadian destinations.

A persistent finding in virtually every study has been the importance of natural and cultural features both in terms of what travelers look for when they plan a trip and activities or places chosen when they actually go to a destination. For example, among those traveling from the United Kingdom, 57 percent indicate that they visited a national park or forest, while 26 percent of those from Japan noted the same thing. Further, when we examine national-park-related travelers in terms of variables like age, we observe at least two themes. Interest in these natural environments cuts across all age groups, although there do appear to be some differences both within and across countries. Further, when we probe about activities done, the age differences become somewhat more apparent. For example, everyone likes to shop, but when we look at outdoor kinds of activities, age becomes more of a discriminator. The number of outdoor activities tends to decrease, the length of time spent doing those activities changes, expenditures associated with the activities also change, and activities are specifically organized and often built in as part of a package arrangement.

Travelers also differ by the types of trips they take. For example, the dominant type for the Japanese is a touring trip, while for those from the United Kingdom, it is visiting friends and relatives. In each instance, there is still a number of these visitors who express interest in natural environments as one of the most important criteria when choosing a place to visit. Another issue is the use of package trips. Japanese travel is dominated by this way of organizing a trip, with almost 70 percent of their long-haul travel being done through trip packages. Other countries exhibit levels of 20 to 30 percent for package tours. Recognizing these differences and/or groups are important in terms of information, communication, service, and possibly influencing where travelers might be enticed to go.

In addition, the activities chosen are rarely those that can be delivered by one person, group, or organization. For example, the list of top five activities might include sightseeing, shopping, sampling local food, a guided excursion, and going to a scenic landmark. The challenge for natural resource organizations becomes how to build a set of linkages or partnerships that can effectively deliver the experiences being sought, particularly for persons from other cultures.

Another key study was the Target Canada focus on the U.S. pleasure travel market conducted in 1989 by Longwoods Travel United States. This was a syndicated survey of a broad range of information objectives based on a methodology of an earlier customized study done for Tourism Canada in 1985. This 1989 study was one of the largest and most comprehensive investigations of the U.S. pleasure traveler, with more than eighteen thousand respondents participating. The survey results suggested that Americans took a total of 959 million person trips for pleasure in 1989, and that 97 percent of all travel was domestic. About 1 percent of the travel reported involved Canadian destinations. In the two stages of the survey, a major focus of the work centered on in-depth investigations of specific trip types. The data gathered are very wide ranging, including many of the same factors gathered in the international survey noted above. Perhaps more interesting has been the recent attempt to identify those individuals who might have the highest propensity to travel and how these might relate to touring, city, and outdoor types of trips.

A final example of recent research is the Canada Outbound Study done by the U.S. Travel and Tourism Administration. The objective of the study was to provide strategic information for use in planning with respect to the rapidly expanding Canadian outbound market. It used the basic design of the international long-haul pleasure travel surveys, but adapted it to the special situation of two adjacent countries and special partner objectives at the state and destination-specific levels. In August and September of 1991, a telephone interview was conducted of 5850 Canadian adults who had traveled outside Canada in the last three years and was followed up by a self-completion mail questionnaire to these same persons. The study showed that the United States continues to be the major drawing card for Canadians, with more than 80 percent of outbound travelers in 1990. In 1990, overnight travel to the United States had surpassed travel between the provinces. Canadians took almost $15 billion out of the country, $7 billion more than was brought in. Same-day travel to the United States has more than doubled in the last ten years. These findings underscore the changes that have been going on in North American travel.

IMPLICATIONS FOR AMENITY RESOURCES

There are probably two key issues that emerge from the investigations of domestic and international travelers. First, there are significant changes occurring at the national, regional, and site levels in the origin, destination, and characteristics of travelers from North America as well as from other parts of the world. These changes imply that there are new and evolving markets that organizations in both the public and private sector will have to address, either independently or as a partnership. In some instances the number of travelers has grown (e.g., Canadian outbound travel to the border states in the United States; overseas travel to Canada and the United States), while in other situations, travel has declined (American travel to Canada). Organizing ways to get travelers to substitute destinations or to choose different travel product opportunities could be a great challenge in the travel industry. Resource organizations have a major role, in that they often provide the settings and infrastructure for travel activities, experiences, and products. Since advertising does influence expectations, cooperative linkages between tourism, marketing, parks, and natural and cultural resources agencies are essential. It is important to reach the international audience in advance to create appropriate expectations for all involved: the visitor or guest, the marketer, and the eventual host, the natural resource destination of choice.

Natural resource agencies in both the United States and Canada have vacillated in terms of defining their roles in relation to tourism and the industry. One could suggest that an agency like the Canadian Park Service has gone from a marketing perspective to a protectionist mode and back to marketing again. The U.S. Forest Service exhibits a pattern similar to this, suggesting an interest in the international visitor in terms of being a customer and even of developing research projects in the organization that address tourism, but then appearing to back away toward their more traditional ecosystem emphasis. The U.S. National Park Service has developed statements about the role of the visitor but has had only limited follow-through, particularly in terms of the international audience.

The Canadian Park Service has probably been most active (although still on a very limited scale) in using tourism data as policy shifts toward marketing have occurred. These data would play a role in looking at service, communication, and operations but have little use in strategic planning, except as background.

However, the oscillations toward and away from marketing probably limit the ability of the agencies to deal most effectively with the tourist industry because the signals are regularly changing. In addition, changes in the domestic North American travel market toward more fragmentation and diversity with higher service requirements also strain resource agencies where attention to downsizing, reinvention, and current policy positions make it less clear how they are expected to proceed.

The creation of partnerships to address the changes in traveler profiles will become a critical issue. In fact, partnerships should abound. First, it is clear that international visitors will create special communication challenges. In addition, for both international and domestic visitors, and for different reasons, there will be demands for special requirements from a nonuniform array of visitors that will stretch the talents of the agencies. We note that many of the activities people pursue will only rarely be within the purview of the agency to provide. One way to contribute to the overall experience is to be certain that some of these opportunities are provided, on site or somewhere close by.

The second issue of concern is the application of the information that is being put together. Most resource organizations are outside the ''loop'' of tourism information, particularly at the international level. In an effort to ''engage the public,'' resource agencies and destinations must raise their levels of communication outside of their organizational boundaries. Private and public tourism groups must reciprocate. While there are a few examples in Canada and the United States where this has occurred, there is still much to be done to structure these ''preengagement meetings.''

Managing for the New Forest Visitor: The Impact of Changing Demographic Variables[1]

Alan W. Ewert

Managing visitors in a resource-based environment such as a national park or forest involves dealing with a constantly changing mix of visitor characteristics, expectations, and opportunities. Accommodating these changes will affect patterns of visitor behavior. It is argued in this chapter that changes in visitor behaviors are predicated on a variety of demographic variables, including population characteristics and other impacting trends.

No longer can the leisure experience be thought of as an episodic event divorced from the broader realities of an individual's life or that of society. It is argued in this chapter that understanding some of the broader social trends impacting our society will lead to the development of better information concerning the management of visitors to our national parks and forest lands.

For example, it has long been held that there is a dichotomy between those recreation visitors coming from an urban environment and those from a more rural environment. That is, the rural-based visitor would be interested in a more primitive recreational experience, while the urban-based visitor would be attracted to developed sites with more amenities (Knopp, 1972). Part of the explanation for this difference is ascribed to variation in the available opportunities and levels of stress held by the urban resident. Recent research, however, has suggested that this rural-urban dichotomy is breaking down, if indeed it ever really existed (Spencer et al., 1991). In its place there is a growing tendency toward a "homogenization" of culture both for society and the recreation visitor, regardless of place of residence.

This is not to say that important differences do not exist; rather, as this chapter will discuss, a number of population factors have a profound and inexorable impact upon many of the attitudes and consequent behaviors of the recreation visitor.

In this chapter, the population variables examined are population growth rate, population age, growth of minority groups, population distribution, diversity of household types, disparity of socioeconomic resources, and available leisure time. This chapter concludes with a brief synthesis of what these trends, in aggregate, suggest for future management needs and challenges.

Underlying this discussion are several assumptions that impact the subsequent conclusions based on demographic and psychometric variables. First, old age and a decreased economic status serve to decrease mobility, which in turn, decreases recreation involvement. Second, an individual's predisposition to visit a park and/or forest is dependent on a number of complex variables, such as desired expectations, intentions to behave, persuasion, and cognitive dissonance (O'Keefe, 1990; Petty et al., 1992).

POPULATION CHARACTERISTICS

Essentially, there are four major patterns in population change that are sufficiently pervasive to affect future events in significant ways. These patterns include

- decreased rates of population growth
- an aging population
- an increasing number and proportion of minority citizens
- a continuing redistribution of the population toward urban areas and toward the southern and western regions of the country

Taken in aggregate, these four patterns represent an overall situation that is fluid and dynamic.

Population Growth Rate

Population growth is determined by three processes: fertility, mortality, and migration (Murdock and Ellis, 1991). As such, population growth is projected to slow substantially, with a maximum size being reached between 2040 and 2050 and declining thereafter. From a service provision perspective, this slowing population growth will require that the performance of private and public sector managers be measured in a variety of ways beyond simply measuring growth rates. For example, growth in the demand for a particular park or forest as a result of increasing population pressure will become a less common and less useful indicator for differentiating levels of performance. Indicators of quality service will become more important in determining whether the needs of the visitor are being met. Parks and forests that are faced with stagnant or decreasing levels of quality service will eventually face decreases in visitation growth rates. This is particularly true in locations faced with heavy competition from other recreation and leisure outlets such as water and theme parks.

In sum, both research and management will be faced with developing a new set of evaluation criteria that measures the success of a particular operation. Growth in participation rates due to increases in overall participation will no longer be the norm. Evaluation of program popularity will need to draw more upon variables such as retention of visitors, length of stay, number of negative and positive comments, and visitor recommendations to others.

Aging of the Population

By the year 2050, it is estimated that nearly 23 percent of the U.S. population will be sixty-five years of age or older (Spencer, 1989). Heinrichs (1991) suggests that not only is the overall population becoming older, but individuals are retiring at an earlier age. Siegel and Taeuber (1986) report that older populations will require more health-related products and a higher level of service intensity. In the use of natural resources, two criteria need to be considered: health of the person and degree of engagement. Doka (1992) points out that older adults who are healthy and seek out new and stimulating environments are the most likely to seek outdoor recreation activities such as park visitation and will present outstanding opportunities for development of travel and hospitality services that cater to the older visitor.

The key for park and forest recreation delivery systems will be how organizations and agencies *accommodate* the older citizen. For example, there will be an increase in the demand for recreational vehicle accommodations with a corresponding decrease in the demand for remote, backcountry opportunities. In sum, management will need to provide for opportunities that are physically active and amenity rich while at the same time being within the capabilities of the older visitor.

Growth of Minority Groups

Juxtaposed to the issue of aging is the growth of minority populations. This is evident both in the differential between an older Anglo population and a younger minority population and in the overall percentage growth in the total population. For example, the percentage of minority members will grow from 25 percent of the overall population of the United States in 1990 to more than 40 percent by 2050. Moreover, minority group members will generally be younger when compared to their Anglo counterparts.

Park and forest recreation delivery systems have evolved as primarily a middle-class Anglo phenomenon. For example, activities such as backcountry hiking and hunting have usually been engaged in by Anglo visitors. With the increasing presence of other cultures and ethnic groups, other activities will become more prevalent. Moreover, recent research has suggested that culture and ethnic background have some influence on the motivations, appeal, and perceived problems of some outdoor recreation areas (Ewert and Pfister, 1991). Future management strategies in park and forest areas will need to be sensitive to the different needs and expectations expressed by minority visitors. Language is one obvious example in which agencies should anticipate an increased need to have multilingual specialists on staff. In addition, differences have been noted in the research literature between Hispanics and Anglos, including (1) time orientation, (2) interpersonal relationships, (3) "power distance" or the level of conformance to authority and regulation, and (4) personal space (Marin and Marin, 1991). Visitor management techniques can and should be sensitive to the ethnic background, cultural ambience, and level of assimilation and acculturation of the individual.

Distribution of Population

Population distribution refers to how a population of an area is distributed as a function of the geographic area and specific site. For example, an individual could live in a large city in the western United States or a small hamlet in the northeast corner of Maine. In general, most developed countries such as the United States have a continuing pattern of increasing population concentration and migration patterns based on climate and economic factors. In the United States there has been an overall migration and increased growth in the southern and western regions relative to the northeastern and midwestern regions. In addition, by 1990, 77.5 percent of the population lived in metropolitan areas, compared to 22.5 percent who lived in nonmetropolitan or more rural environments (Long, 1988). Murdock and Ellis (1991) suggest that this trend will continue but decrease in magnitude.

These trends have several implications for park and forest visitor management. First, the park and forest visitor will increasingly come from an urbanized environment. As such, there will be an emphasis on visiting restorative environments (Kaplan and Kaplan, 1989). Moreover, it can be anticipated that the urbanized visitor will have less knowledge and background in an outdoor environment and may expect or desire to engage in a broader spectrum of recreational activities (Ewert, 1989).

Lessinger (1987) believes that one important aspect of population distribution will be the movement of the population to wildland-urban interface lands, those lands near large urban environments (Ewert et al., 1993). According to Garreau (1991), this movement can already be seen in the development of "edge cities." If true, this microlevel migration pattern will promulgate intense pressure on many recreation and natural resource systems by creating adjacent development and increased impacts on heretofore remote areas.

In sum, from a management perspective, two factors become apparent when considering the distribution of the population. First, where the majority of our citizens live will have a great influence on what they want to do and where they will go with respect to park and forest visitation (Kelly, 1989). Second, migration patterns will have a tremendous, often negative, influence on both the management of park and recreation areas (e.g., overcrowding) and the impacts to the natural resource base (e.g., air and water pollution).

PATTERNS OF CONTINUING IMPORTANCE

In addition to these four major patterns of population characteristics there are several other patterns that will affect the visitation rates and behaviors of the public to park and forest areas. These patterns include (1) the continuing diversity of household types, (2) a continuing disparity in socioeconomic resources, and (3) changing patterns in available leisure time.

Diversity of Household Types

Because of the aging population base, the rate of household formation will slow. Social and economic forces, however, will continue to create highly diverse households, often characterized by one-parent families or diverse forms of union (Bumpass and Sweet, 1989). Of even greater interest, from a park and forest management perspective, are the associated changes in lifestyle. The research is consistent in suggesting that different lifestyles and developmental periods of life (e.g., childhood, late adulthood) often lead to different expectations and demands for leisure activities (Kelly, 1989). Overall, the census data predict a reduction in "traditional family" units, a growth in units other than two-parent unions, a growth in multi-income family units with a corresponding decrease in available leisure time common to other family members, and smaller sized consuming units, i.e., smaller families (Fosler et al., 1990; Marcin, 1993). Ultimately, this implies that the "traditional" two-week vacation is rapidly becoming irrelevant to much of our population and will be replaced by visits to recreation areas that are shorter, that take place on the weekends, and that involve less planning and travel time.

Disparity in Socioeconomic Resources

Disparities in socioeconomic resources for different populations have for decades led to differentials in available opportunities, income, occupational mobility, and educational attainment (Farley and Allen, 1987). Census data suggest that while overall income levels are relatively high, poverty rates also remain high and are growing for some minority and other economic groups. From a leisure and recreation perspective, these socioeconomic disparities can impact the visitation of parks and recreation through visitation rates and behavior patterns. In some cases, visitation to local parks and forests will increase due to the relatively lower cost and fewer planning requirements. Local parks and forests may become destination points primarily because they are "close, accessible, and cheap." In this type of scenario, it can be expected that visitation to more remote sites will decrease, while significant increases will be seen in urban and municipal parks as well as areas that are located in the wildland-urban interface (Ewert et al. 1993).

Available Leisure Time

A Louis Harris survey in 1987 reported a decline in leisure time for adults from 26.2 hours per week in 1973 to 16.6 hours per week in 1987. This drop was due in part to a shift toward a service economy, more salaried workers (who generally work longer hours), longer commuting times, going to school, and an increase in the number of women working outside the home. When broken down by gender, men averaged 20.3 hours of leisure time, and women had only 15.6 hours. From a recreation perspective, money and time are now the two great limiting factors for participation, with time beginning to supplant money as the most important variable. The time variable, like the economic resource variable, results in leisure activities that are "shorter, closer to home, and cheaper." Scott (1993) also reports under a situation of time scarcity, individuals will seek to maximize their time by engaging in goods-intensive behavior (e.g., power boats instead of canoes), simultaneous activities (e.g., reading some work-related material while fishing), and becoming more self-directed (e.g., seeking a variety of options to choose from).

CONCLUDING COMMENTS

Demographic variables such as population characteristics and socioeconomic considerations play important roles in visitor behaviors and rates of participation at park and forest areas. It seems reasonable to expect that the traditional two-week vacation to a remote park or forest area for camping and fishing with the family may be diminishing in importance. Whether for good or bad, this experience will be replaced by something that is shorter in length, more inclusive of other people, closer to home, and more demanding in terms of different activities to pursue. The recreational experience in a park or forest setting will increasingly be more pluralistic, culturally diverse, and less hinged on past experience and knowledge levels. For example, most future visitors will not have a belief system that is built around childhood experiences of camping with the family, organized camps, or visitation to remote and pristine national parks and forests. As a result, issues and behaviors such as littering, cutting vegetation, noise pollution, and intrusion in other people's space will have to be relearned, if indeed they

Table 1. Demographic Change and Implications for Natural Resource Management

Demographic changes	Implications
Decreasing growth rate for population	Lack of reliance on popularity New evaluation systems for determining success
Aging of population	Need to accommodate older visitor. Greater demand for amenities Demand for activities that are less physically demanding Greater demand for "front-country," less demand for backcountry opportunities
Growth of minority groups	Potential change in activity mix Need for managers to have more comprehensive and wide-ranging communication training
Population distribution	Movement to wildland-urban interface lands Greater percentage of "urbanized" visitors
Less available leisure time	More reliance on technology Visitation that is shorter, closer to home, and cheaper Demand by visitors for more choice in activities

are learned at all. Ultimately, what this implies is that agencies and organizations will be forced to spend more effort on educating the new visitor into the basics of outdoor environmental ethics and behaviors.

ENDNOTE

1. Submitted to: Proceedings—Second Canada/U.S. Workshop on Visitor Management in Parks, Forests and Protected areas, Madison, Wisconsin, May 13–16, 1992

REFERENCES

Bumpass, L., and J. Sweet. 1989. National estimates of cohabitation. *Demography* 26:615–625.

Doka, K. (1992. When gray is golden: Business in an aging America. *The Futurist* 26(4): 16–20.

Ewert, A. 1989. Wildland recreation: Bone games of the past or salvation of the future? *Trends* 26(3): 14–18.

Ewert, A., D. Chavez, and A. Magill. 1993. *Culture, conflict, and communication in the wildland-urban interface.* Boulder: Westview Press.

Ewert, A., and R. Pfister. 1991. Cross-cultural land ethics: Motivations, appealing attributes and problems. *Transactions of the North American and National Resource Conference,* 56:146–151.

Farley, R., and W. Allen. 1987. *The color line and the quality of life in America.* New York: Russell Sage Foundation.

Fosler, R., W. Alonso, J. Meyer, and R. Kern. 1990. *Demographic change and the American future.* Pittsburgh: University of Pittsburgh Press.

Garreau, J. 1991. *Edge city: Life on the new frontier.* New York: Doubleday.

Heinrichs, J. 1991. The future of fun. *American Forests* (March/April): 21–24, 73–74.

Kaplan, R., and S. Kaplan. 1989. *The experience of nature: A psychological perspective.* New York: Cambridge University Press.

Kelly, J. 1989. Leisure behaviors and styles: Social, economic, and cultural factors. In *Understanding leisure and recreation: Mapping the past, charting the future,* edited by E. Jackson and T. Burton, 89–112. State College, Pennsylvania: Venture Publishing.

Knopp, T. 1972. Environmental determinants of recreation behavior. *Journal of Leisure Research* 4:129–138.

Lessinger, J. 1987. The emerging region of opportunity. *American Demographics* 9(6): 33–37, 66–68.

Long, L. 1988. *Migration and residential mobility in the United States.* New York: Russell Sage Foundation.

Marcin, T. 1993. Demographic change: Implications for forest management. *Journal of Forestry* 91(11): 39–45.

Marin, G., and B. Marin. 1991. *Research with Hispanic populations.* Newbury Park, California: Sage Publications.

Murdock, S., and D. Ellis. 1991. *Applied demography: An introduction to basic concepts, methods, and data.* Boulder: Westview Press.

O'Keefe, D. 1990. *Persuasion: Theory and research.* Newbury Park, California: Sage Publications.

Petty, R., S. McMichael, and L. Brannon. 1992. The Elaboration Likelihood Model of persuasion: Applications in recreation and tourism. In *Influencing human behavior,* edited by M. Manfredo, 77–102. Champaign, Illinois: Sagamore Publishing.

Scott, D. 1993. Time scarcity and its implications for leisure behavior and leisure delivery. *Journal of Park and Recreation Administration* 11(3): 51–60.

Siegel, J., and C. Taeuber. 1986. Demographic perspectives on the long-lived society. *Daedalus* 115:77–177.

Spencer, G. 1989. Projections of the population of the United States, by age, sex, and race: 1983 to 2080. *Current Population Reports.* Series P-25, No. 1018. U.S. Bureau of the Census, Washington, D.C.: U.S. Government Printing Office.

Spencer, R., J. Kelly, and J. Van Es. 1991. Residence and orientations toward solitude. *Leisure Sciences* 14:69–78.

Chapter 4

Mobilizing for Environmental Justice in Communities of Color: An Emerging Profile of People of Color Environmental Groups

Dorceta E. Taylor

In recent years, people of color[1] environmental groups have brought the issues of environmental racism, environmental equity, environmental justice, environmental blackmail, and toxic terrorism to the forefront of the environmental debate. Until people of color made the usage of these terms commonplace in environmental circles, these expressions, the concepts they embody, and the questions arising from them were not used, explored, or asked by traditional and well-established environmental groups, deep ecologists, social ecologists, bioregionalists, ecofeminists, or greens. Environmental activists (even the more radical ones or those who were critical of traditional environmental activism) ignored or paid little attention to the processes, practices, and policies that led to grave inequities, to charges of environmental racism, and to a call for environmental justice. For a long time, environmentalists did not recognize that certain issues, processes, policies, and activities had disproportionate negative impacts on communities of color; if they were aware of the impacts, they did not pay any attention to them. This occurred because many in the environmental movement did not perceive and define some of the issues affecting communities of color as environmental issues, they did not consider people of color part of the constituency they served, or they did not see themselves engaging in environmental dialogues and struggles with such communities. If and when they considered people of color, it was as an afterthought deserving only of marginal consideration. In addition, many environmentalists were too concerned with other environmental issues to move issues primarily affecting people of color to the top of their agendas.[2]

People of color have been at the forefront of the struggle to bring attention to the issues that are devastating minority communities—issues like hazardous waste disposal, exposure to toxins, occupational health and safety, housing, pollution, environmental contamination, lead poisoning, pesticide poisoning, cancers, and other health issues. Their communities, some of the most degraded environments in this country, are the repositories of the by-products of capitalist production and excessive consumption. As a result, they have been in the vanguard of the struggle for environmental justice. Yet, despite the rapid mobilization in communities of color, the formation of numerous environmental groups of color and an impressive record of environmental victories, very little is known and written about these groups in mainstream academic and policy-making communities. One consequence of this is the persistence of the stereotype that people of color are not interested in environmental issues (at least not as interested as whites) and are not likely to mobilize around such issues.

This paper will focus on a small aspect of people of color environmental activism. It will look at the mobilization of people of color around environmental issues in the broader context of the history of the environmental movement. It will analyze the similarities and differences between the larger environmental movement and people of color environmental groups, and it will also discuss the emergence of the environmental justice movement—the sector of the environmental movement that people of color are most active in.

WHITES AND ENVIRONMENTAL ACTIVISM

During the middle to late 1800s, the writings, drawings, and exhortations of activists and visionaries like John J. Audubon, Ralph Waldo Emerson, George Perkins Marsh, Henry David Thoreau, John Muir, and Gifford Pinchot laid the groundwork for the emergence of the environmental movement. By the 1870s, environmental groups like the Appalachian Mountain Club were formed, and the transition from the premovement era (dominated by individual enthusiasts and scientific and technical professionals) to mass movement began to take shape. Although some researchers[3] argue that the environmental movement did not begin until after the publication of *Silent Spring*,[4] this chapter contends that the environmental movement began in earnest around the time of Hetch Hetchy and that environmental activism can be divided into four phases, two of which (the premovement and post–Hetch Hetchy eras) came before *Silent Spring*[5]. As subsequent discussion will show, to argue that the environmental movement did not begin until the 1960s would require one to discount the existence of large numbers of environmental groups (accounting for a significant percentage of the environmental groups in existence today); their lobbying, legislative, policymaking, and coordinating efforts; the significant influence of the environmental agenda they established and ideology they espoused; the collective identity they developed; and the large membership they had.

The Hetch Hetchy affair thrust environmental issues onto the public stage.[6] For the first time, citizens who were not a part of the small elite group of preservationists, conservationists, and outdoor enthusiasts got involved in environmental debate. They did so through letter-writing campaigns, newspaper articles, and public debates.[7] During the debate, the conservationists in particular mobilized political allies at the highest levels in the federal and state governments, technical assistance, the media, and citizens, who wrote letters. Both sides struggled with ways of *framing their protests*.[8] Both sides took existing issues and framed them in ways that expanded, amplified, and transformed the issues, all the time being careful not to frame the issues in such a way that they were too far ahead of their followers. Despite the media campaigns and public participation, San Francisco was granted permission to dam Hetch Hetchy in 1913.

It is not surprising that a movement sprang out of one of the first major public environmental controversies. Listings in the 1993 and 1994 *Conservation Directory* and the 1992 *Gale Environmental Sourcebook*[9] show that by the turn of the century, many of the elements of what would become the environmental movement were already in place. There were numerous national, regional, and state organizations already in existence. In addition, major environmental and natural history institutions like the Smithsonian had been around for decades, and many professional environmental associations and a farmer's association existed. Prior to Hetch Hetchy, these institutions and organizations did not coordinate their activities extensively, but Hetch Hetchy focused attention on one issue and that allowed for the necessary coordination and communication to take place. Because of Hetch Hetchy, people became very aware of the need to work toward the preservation of natural areas.

What occurred at this place of the movement's history is very much in line with the postulations of McAdam (1982) and Friedman and McAdam (1992), who argue that new movements usually emerge out of existing movements and social institutions, remaining dependent on the preexisting movements and institutions for some time. This step is crucial because of the resources and the collective identity (status) associated with the older movements/institutions. However, for fledgling movements to be successful, they have to eventually disentangle themselves from preexisting movements and institutions and forge a distinct identity. In the case of the environmental movement, it distanced itself from the animal rights and the zoo movements and redefined itself so as not to be seen purely as a parks or playground movement.[10] Environmentalists also had to shed the image of being merely birdwatchers, mountaineers, hunters, and anglers indulging their own interests with little concern for the public good.

Table 1 Formation of Predominantly White Environmental Groups, 1845–1994

Year founded	Frequency	Percentage
1845–1875	14	1.4
1876–1899	30	2.9
1900–1909	24	2.4
1910–1919	24	2.4
1920–1929	30	2.9
1930–1939	32	3.1
1940–1949	39	3.8
1950–1959	77	7.6
Cumulative percent		26.5 (n = 270)
1960–1969	170	16.7
1970–1979	289	28.4
1980–1989	270	26.5
1990–1994	19	1.9
Cumulative percent		73.5 (n = 748)

Compiled from the *Conservation Directory* (National Wildlife Federation, 1993, 1994) and the 1992 *Gale Environmental Sourcebook* (Hill and Picirelli, 1992).

Table 1 shows that during and immediately after Hetch Hetchy, a number of new environmental organizations were formed.[11] The fledgling movement had a very narrow agenda—the groups focused on wildlife and forest preservation, hunting, birdwatching, fishing, hiking, and mountaineering. By the 1930s, the movement began to stagnate—the growth of the movement slowed; the political activities and issues being tackled failed to capture the imagination of large sectors of the population. This period of malaise continued through the 1940s.[12]

Compared to later periods in the movement, this era (the post–Hetch Hetchy era) was also marked by a great deal of accord on the goals, agenda, and role of the movement. This occurred despite the fact that the movement's agenda broadened somewhat during this era. Though there were disagreements between some of the most influential groups over the nature of their alliances with the gun manufacturers,[13] the groups being formed were, for the most part, *consensus* organizations,[14] and their disagreements were not about deep ideological rifts between the groups. Though some, like the Izaak Walton League, were recruiting new members effectively, they did not recruit as effectively as environmental groups that emerged in later periods of the movement.

Consensus movements refer to kinds of social mobilizations that enjoy broad attitudinal support (if not necessarily other kinds of public support like monetary or membership support) and meet with little or no organized opposition. Usually, more than 80 percent of the population generally supports the goals of such movements, which tend to describe themselves as "nonpolitical," "educational," "nonpartisan," or "humanitarian," and tend to avoid conflictual situations.[15] In addition to their ideological positions, consensus movements can also be characterized by the social infrastructures and institutional resources that shape their emergence and (in some instances) support their continued growth.[16] Consensus movements can be contrasted with *conflict* movements, which are characterized as social movements that attempt to restructure the social order, change public policies, and shift the balance of power among groups. Such mobilizations usually encounter organized opposition.[17]

The environmental movement began to emerge from its stagnation during the 1950s—the number of groups formed during this decade was almost double that of the previous decade. As the rapid increase in the formation of new groups continued during the 1960s, the number of groups formed in this decade more than doubled that in the 1950s. This decade ushered in a new era of environmentalism—the publication of Rachel Carson's *Silent Spring* in 1962[18] led to an immense public outcry over pollution that launched the birth of the modern environmental movement. That outrage sparked a mass mobilization drive that resulted in cleaner air, rivers, and lakes for many Americans. For the first time, radical environmental groups were formed, and toward the end of the decade, many youths (students, peace activists, antiwar protesters, civil rights activists, antinuclear activists) joined the movement. Prior to this time period, the environmental organizations could best be described as primarily *reformist* in orientation.[19] They sought to work closely with government and industry, seeking only marginal or incremental changes (reforms) in the system. They did not seek to change the status quo very much.

During this period—the post-Carson era (1960–1979)—radical, *conflict* groups like Greenpeace were formed. These conflict groups mastered the art of framing protests effectively, used injustice frames, and redefined environmental activism. The agenda of the movement was broadened dramatically; issues like energy, toxics, and pollution became the concern of many groups. Several new components of the movement were added during this period. Submovements like ecofeminism also took hold during this time period. These *submovements* were more than just groups working on issues; they were movements that shared some general goals (like halting the wanton destruction of natural resources) with the mainstream environmental movement but differed significantly from the mainstream and from each other in more specific goals, political and ideological orientation, and strategies for accomplishing their goals. They issued ideological critiques of the government, industry, and the existing environmental movement; they made forecasts about the future of humans and the earth; and they offered alternative visions of the future and of environmental action.

The surge in activism carried over to the 1970s—the period in which most of the environmental groups were formed. Almost twice as many groups were formed in this decade as in the 1960s. However, by the 1980s (the post-Three Mile Island/Love Canal era),[20] many of the radical environmental groups of the sixties and seventies adopted reform agendas, distanced themselves from grassroots (conflict) politics, established offices in Washington, D.C., and focused their efforts on lobbying, filing lawsuits, producing technical information on the issues of interest to them, and recruiting new members through mail campaigns.

For most of this century, environmentalists have been effective at mobilizing the necessary resources, framing the issues, and controlling the discourse in ways that capture public attention.[21] Despite the successful transformation into an influential mass movement, the environmental movement has never had a strong base of support among the poor, the working class, and people of color. However, environmental accidents at the Three Mile Island nuclear plant in Pennsylvania and toxic contamination at sites like Love Canal (New York) and Times Beach (Missouri) provided the catalyst for a new era of grassroots environmental activism focused on toxics and other environmental hazards. Many working-class whites and people of color increased their activism or became active in environmental campaigns for the first time (see Table 2).

Though consensus reform organizations continue to emerge during the 1980s and 1990s, it is the emergence of many persistent, radical grassroots organizations that have profoundly changed the nature of the environmental movement once again. While scholars who study political participation in the environmental movement did not predict the increased and sustained mobilization of working-class activists, the growth of the grassroots sector has surprised many.[22] For instance, Citizens Clearinghouse for Hazardous Wastes (CCHW) worked with about two thousand groups in 1989 but now works with over seven thousand grassroots groups nationwide.[23]

One key factor explaining the increased growth and sustained activism is mobilization around toxics. Mazmanian and Morrell[24] offer an explanation of why the toxics issue has not followed the typical issue-attention cycle postulated by Anthony Downs.[25] If the cycle were adhered to, the issue of toxic contamination would have receded in importance in the public's mind since it first gained

Table 2 Time Periods during Which People of Color Environmental Groups Were Formed

Time period in which organization was formed	Number of organizations formed (n = 331)	Percent of organization
1845–1959	35	10.6
1960–1969	22	6.6
1970–1979	49	14.8
Total	106	32
1980–1989	160	48.3
1990–1994	65	19.6
Total	225	67.9

Compiled from the *People of Color Environmental Groups Directory* (Bullard 1992b, 1994) and telephone interviews with environmental groups.

national attention in 1978 if media coverage had waned, if the quick-fix techno-logical solutions offered by bureaucrats had worked, and if other problems had emerged to eclipse the resonance that toxics had in people's minds. The toxics issue has remained in the news and has continued to be a mobilizing factor because what people thought was the worst-case scenario when the story first captured national attention in the late 1970s turned out to be just the tip of the iceberg. The constant parade of new discoveries of toxic contamination (each new one seeming worse than the one that preceded it) has kept the issue at the top of the agenda of many activists and foremost in the minds of many citizens. In addition, many huge corporations—some long-time providers of jobs and supporters of local civic organizations and events—were found to be the source of the contamination. Many people felt that the trust or social compact between host communities and corporations was broken. People were further dismayed to find that the government, policy makers, scientists, and other experts could not offer ready solutions or any solutions at all. In many instances, nothing was done while the government, business, and scientists remained bogged down in long delays due to legal or technical skirmishes. In some cases, people discovered there were no safeguards to protect them from the whims of industry. The state lost legitimacy,[26] and people, feeling they had been deserted by the state and mistreated by big business, mobilized to shift the balance of power and to ameliorate the situation. Some of these radical groups participate in the environ-mental justice movement, one of the submovements that emerged in the 1980s.

THE EMERGENCE OF PEOPLE OF COLOR ENVIRONMENTAL GROUPS

While there are scores of books and articles documenting the history of environ-mental activism and the environmental movement, these are histories and docu-mentations of white environmental activism and the white environmental groups that comprise most of the movement. For example, some of the environmental justice groups have been in existence since the mid-1800s and actually predate the formation of the early predominantly white environmental groups,[27] yet they are not recognized by the mainstream environmental movement or by scholars who study the environmental movement. Native Americans forced onto reserva-tions formed some of the earliest organized environmental groups on record, yet these groups and the ones that were formed later have not been incorporated in the environmental history.

To illustrate this point further, until a few years agao most environmental groups of color went unrecognized by the largely white environmental movement. For example, in the editions of the *Conservation Directory* analyzed for this chap-ter (National Wildlife Federation, 1993, 1994), only 5 of the 331 groups listed in the *People of Color Environmental Groups Directory* (Bullard 1992b, 1994) were listed.[28] In fact, more Canadian and international groups were listed in the directo-ries than American environmental groups of color.

The Birth of an Environmental Justice Activist: Case of the "Toxic Doughnut"

Hazel Johnson's story is typical of the way many activists become involved in the environmental justice movement. Hazel lives in Altgeld Gardens, a predominantly black housing project on Chicago's Far South Side. She refers to the neighborhood of ten thousand residents as a "Toxic Doughnut" because homes are encircled by landfills, factories, and other industrial sites that emit toxic and/or noxious fumes. West of the Doughnut, the coke ovens of Acme Steel discharge benzene into the air, to the south is Dolton's municipal landfill, to the east is Waste Management's CID Corp. landfill, and to the north near 130th Street and Stony Island Avenue lie beds of sewage sludge at the Chicago Metropolitan Water Reclamation District facility. There are fifty abandoned hazardous dump sites within a six-mile radius of this neighborhood. The toxic stew around the Doughnut is so potent that Illinois inspectors aborted an expedition in one of the dumping lagoons when their boat began to disintegrate.[29]

Illness was common in the area, but it was not until her husband died of lung cancer and other family and friends became ill that she started to wonder if there was an environmental link to the deaths and illnesses. She conducted an informal health survey of one thousand of her neighbors and was astounded at the number of cancers, birth deformities, premature deaths, skin rashes, eye irritation, and respiratory illnesses that were reported. She and the environmental justice group she founded, People for Community Recovery (PCR), contacted the City of Chicago about the findings and urged them to investigate the illnesses. The city conducted a controversial study that found high rates of cancers among African Americans on Chicago's South Side, but did not investigate whether the rate was higher than that found in African Americans elsewhere or whether the health effects were in any way related to the exposure to toxins in the area. Dissatisfied with such inconclusive findings PCR commissioned their own study and also requested the Agency for Toxic Substances and Disease Registry (ATSDR) to do a health study. The ATSDR agreed to do a health assessment of the area. In the mean time, members of PCR, suspecting that some of the health problems might be caused by contaminated sulfur-smelling well water, lobbied for and obtained municipal water hookup. Additionally, PCR, on discovering that Waste Management wanted to expand its CID landfill site, staged a series of protests (with the help of Greenpeace) that blocked the expansion. PCR, Greenpeace, and other local groups also challenged Chemical Waste Management Inc. on its practices at its South Stony Island Avenue incinerator. An investigation into the groups' claims by the Illinois Environmental Protection Agency found numerous violations, including disabled stack monitors and the storage of eighty thousand gallons of waste in excess of what the law allowed. The company paid $7.2 million in penalties between 1980 and 1991, and $500,000 to an environmental scholarship fund.[30]

Key Movements and Institutions

The history of people of color environmental activism is not well documented. Although the history of political activism of African Americans, Latinos, Native Americans, and Asian Americans began on divergent paths, they have had commonalities—the struggle to end blatant and vicious discrimination and genocide. The common struggle for justice has drawn these groups together, with the emergence of the environmental justice movement and with strong alliances being formed among environmental groups of color nationwide. The history of people of color environmental activism is increasingly becoming a history common to all these groups. Although the details are sketchy, there is enough information to provide a profile of these groups. People of color environmental groups have emerged out of some key movements and institutions (see Box 1).

Box 1   Key Movements and Institutions

1. Native American Tribal Groups; Sovereignty and Traditional Rights
2. The Civil Rights Movement
3. The Farm Worker Movement; Pesticides and Worker Rights
4. The Occupational Health and Safety Movement; Worker Rights
5. Human Rights Movement
6. Religious Institutions
7. Social Services Institutions
8. Educational Institutions

Native American groups struggling with the erosion of cultural values and treaty rights have used these issues to call attention to hazards of living on reservations. Native Americans like the Navajos living near Rio Puerco, New Mexico, face increased health risks from the numerous uranium mines around them. Navajos face an increased risk of cancer from drinking water tainted with radioactive substances or from eating animals grazing in areas contaminated by the mining operations. Consequently, the Rio Puerco Navajos have developed a strong environmental justice agenda focused on educating residents about environmental conditions, bringing clean water to the area, and resolving land rights issues in the area.[31] Similarly, Native Americans on the St. Regis Mohawk Reservation in New York face severe health hazards from exposure to polychlorinated biphenols (PCB).[32]

Many African American groups and/or leaders have their roots in the Civil Rights movement. Groups like the Gulf Coast Tenants Association, which was founded to help to improve housing conditions of blacks in the South, have taken on strong environmental justice and environmental education agendas as they are faced by the reality of the health risks to communities of color in the South. Working in and around "Cancer Alley," the ninety-mile stretch running along the Mississippi River from Baton Rouge to New Orleans and home to about one-fourth of the chemical manufacturing plants in the United States,[33] members of the Gulf Coast Tenants Association are in constant communication with communities in which chemical spills and "accidental" releases of toxins into the environment are more routine than accidental. These communities have high rates of cancers, birth defects, miscarriages, infant mortality, and respiratory illnesses.

While African American and Native American environmental justice groups have included labor issues in some of their environmental justice struggles, Latinos in the farm worker movement make this linkage one of their key organizing tools. Through their participation in the United Farm Workers Union, farm workers in California and other parts of the South and West have launched successful grape boycotts and focused the nation's attention on the harmful effects of pesticides. They have documented their illnesses, which include acute and chronic pesticide poisoning, death, infertility, birth defects and miscarriages, skin rashes and eye irritation, and respiratory infections. They have also documented clusters of childhood leukemia in areas such as McFarland, California.[34]

Similarly, Asian Americans concerned about immigrant rights and working conditions in the computer and garment industries have also formed environmental justice groups. Asian women working in the computer chip factories in Silicon Valley tell of birth defects, miscarriages, and disabling illnesses and injuries that occur because of the chemicals they are exposed to in their line of work. They are involved in groups like the Asian Women's Advocates and the Santa Clara Center for Occupational Safety and Health.[35] Other environmental justice groups like Southwest Organizing Project (SWOP) have been in negotiations with computer manufacturers in the Southwest on ways of mitigating environmental damage.[36]

The occupational health and safety movement, a predominantly working-class movement, has also been key to linking the struggles and concerns around improving working conditions with environmental justice issues. All over the country, Committees on Occupational Safety and Health (COSH) have helped to coordinate the efforts to reduce workplace hazards in each state.

Chronology of Key Historical Events

In addition, some pivotal historical events have greatly influenced the participation of people of color in the environmental justice movement. Some of these key events include the contamination of the land and waterways of Triana, Alabama, a predominantly black community heavily reliant on subsistence fishing, with DDT in 1981. Tests showed that residents had in their blood streams some of the highest levels of DDT ever recorded.[37] Another event in 1982, namely, the building an enormous landfill (called the "Cadillac of landfills" by Chemical Waste Management) in predominantly black Emelle, Alabama, set off a series of protest actions. Those demonstrations spurred the commissioning of the 1983 U.S. Government Accounting Office study of landfill siting in the southeastern United States. Though small in scope, this study indicated that race is an important factor in siting decisions.[38]

A 1985 urban environment conference was also critical, in that it identified many of the environmental hazards communities faced and it linked health with worker safety and environmental degradation.[39] Two years later, the Commission for Racial Justice (1987) of the United Church of Christ released its study on toxic waste and race. The study documented what by now many communities had already begun to suspect, namely, that communities of color were disproportionately exposed to environmental hazards.

Two critical events occurred in 1990—just in time for the Earth Day celebrations. A group of scholars and policy makers (most of them people of color) met at the University of Michigan to analyze and critique the state of the research on people of color and environmental hazards, to establish better communications (between communities of color, the academic community, the government and policy-making communities, and traditional environmental communities), and to hold government agencies (particularly, the Environmental Protection Agency (EPA)) accountable for the impacts of their policies and actions in communities of color.[40] Two outgrowths of the Michigan conference were the call for a grassroots leadership summit and for the EPA to establish an environmental justice (formerly environmental equity) working group. In 1992, the agency put out its first report on environmental justice.

The second 1990 event took the form of a letter addressed to the Big Ten (renamed Green Group) environmental organizations that was published in the *New York Times* and signed by leaders of environmental justice groups. The letter, which embarrassed most in traditional environmental circles, put the issue of the "whiteness of the green movement" on the environmental agenda in a public way. Most of these organizations either scrambled to demonstrate that they had a few people of color in their membership or on their staff or board, or professed that they had not noticed that their organizations were all—or mostly—white. Planning for the Earth Day festivities was overshadowed by awkward excuses and clumsy attempts to diversify (i.e., break the color barrier) these organizations and events.

By 1991, when over six hundred delegates were invited to the First People of Color Environmental Leadership Summit, activists and community leaders were ready to establish national and regional networks to coordinate their efforts. The conference served to help people understand the scope of the problem, outline a strategy for shifting national policy, identify resources, establish contacts, develop the principles of environmental justice, and renew their commitment to reduce hazards in communities of color. The goal of strengthening regional networks was furthered in 1992, when thousands of people met for the Southern Organizing Conference Summit. In 1994, over a thousand delegates attended one of the first of the research and policy summits on health and environmental justice.

REGIONAL VARIATIONS IN MOBILIZATION

Although the environmental movement emerged as a successful mass movement by the 1970s, the participation of people of color in the movement was limited.[41] Since then, there has been much change in the attitudes of people of color towards environmental activism. A study of 1,462 predominantly white environmental groups found that the peak period of formation of new groups occurred in the 1970s (see Table 1). In comparison, the peak period for the 331 people of color groups occurred in the 1980s (see Table 2). In addition, the number of people of color groups doubled between 1960s and the 1970s. More than three times as many groups were formed in the 1980s than in the previous decade.

The rate at which people of color environmental groups have been formed has also varied regionally. People of color groups differ from predominantly white environmental groups in the regions of the country in which they are most active. Between 1845 and 1959, the Midwest and the Southwest were the two regions in which most of the people of color environmental groups were being formed; however, during the 1960s there was a shift to the Southeast. During the 1980s the Southwest was the most active region, followed closely by the South. The Pacific Northwest and the Midwest had the lowest levels of activism since the 1980s. This pattern can be partially explained by fact that people of color groups grew out of existing social, political, educational, and religious institutions; the social structure; and existing movements.[42] It is not surprising, therefore, that during the sixties the rate of formation of groups that eventually took on an environmental agenda in the South increased because of the Civil Rights Movement (Table 3). During the 1980s, there was also an increased rate of formation of groups in the Southwest, due in part to the mobilizations around pesticides and worker rights by groups like the United Farm Workers. The lower rates at which new groups are formed in the Pacific Northwest is partly due to the fact that a significant amount of the people of color environmental activism

Table 3 Region and the Percentage of Environmental Groups that Originally Started as Environmental Groups

Region	Number of groups	Percent that did not start as environ- mental group	Percent starting as environmental group
Nationwide	330	61.2	38.8
Northeast/Mid-Atlantic	66	50.0	50.0
South/Southeast	93	54.8	45.2
Midwest	43	74.4	25.6
Southwest/West	103	64.1	35.9
Pacific Northwest	25	80.0	20.0

Compiled from the *People of Color Environmental Groups Directory* (Bullard, 1992b, 1994) and telephone interviews with environmental groups.

is occurring among Native American tribal groups. Rather than starting small, local organizations, activists join with existing Native American tribes. Forty-six percent of the people of color environmental justice groups in this region are tribal groups that have been in existence for a long time.

By contrast, predominantly white environmental groups have always flourished in the Northeast—the region with the highest rate of new group formation from 1845 to the present. This differs from people of color groups, which arise out of a variety of social and political institutions and have growth spurts in various regions of the country, depending on conditions in these regions and the social welfare of the people living there. However, the predominantly white environmental movement was born among the urban elites of the Northeast and has had strong support (particularly by the upper and middle class) in this region. As articulated and practiced by these elites from the mid–1800s to the mid–1900s, environmentalism was divorced from the social, economic, and political welfare of the urban and rural poor and people of color. This ideological hegemony was not seriously challenged till the 1960s, and then only in ways that brought marginal benefits to people of color. It was not until the 1980s with the increased activism of people of color and the emergence of the environmental justice movement that a wider range of concerns was introduced into the environmental debate.

Two time periods are of particular interest. The first, 1876–1899, is the transition period from individual enthusiasts and activists encouraging people to care for the environment to the formation of environmental groups like the Appalachian Mountain Club and the Audubon Society. This was a period in which the level of activism in the Northeast was significantly higher than that of any other region of the country. A similar pattern emerges from 1960 to 1989, when there were significantly higher levels of new group formation in the Northeast than in any other region. Not only were there many new groups desirous of locating their headquarters close to the governmental and financial centers of the country, but groups that were not previously based in the Northeast established themselves there. Because people of color environmental justice groups want to maintain direct contact with local, grassroots groups, they have resisted the trend to locate their headquarters in the Washington, D.C., or New York areas.

In analyzing the formation of predominantly white organizations, it should also be noted that the South and Southwest have had much lower levels of new group formation than the Northeast. This is in stark contrast to people of color environmental justice groups that are more likely to be formed in these two regions than in any other part of the country. As was the case in people of color groups, the Pacific Northwest had one of the lowest rates of new group formation.

MULTIFACETED GROUPS

Another striking difference between the predominantly white and the people of color environmental groups is that throughout their history, predominantly white environmental groups originate as environmental groups or arise from preexisting environmentally related institutions; they focus solely on environmental issues or on a nexus issue closely related to the environment. In contrast, people of color environmental groups tend to be multiple-purpose groups focused on environmental issues in conjunction with other issues like sovereignty, economic development, health and welfare, and civil and human rights.[43] For instance, some of the people of color groups formed in the premovement and post-Hetch Hetchy eras were Native American tribes, but they were also groups with explicit land, water, and natural resource conservation and preservation policies and strategies.

Sixty-one percent of the groups listed in the *People of Color Environment Groups Directory* (Bullard, 1992b, 1994) were not founded as environmental groups; they started as other kinds of social justice groups that adopted an environmental agenda. This percentage varies by region (see Table 3). The Northeast has the highest percentage of groups (50 percent) that started out as environmental groups; this is twice as high as the percentages in the Midwest and 2.5 times as high as the percentage in the Pacific Northwest. Forty-five percent of the groups in the South and 36 percent of those in the Southwest started out as environmental groups. This can be partially explained by examining the link between the time period in which the groups were formed and whether they started out as environmental groups (Table 4). In general, the later the time period in which the group was formed, the more likely it is that the group started out as an environmental group. While 11 percent of the groups

Table 4 Time Period Founded and the Percentage of Environmental Groups That Originally Started as Environmental Groups

Time period Year founded	Number of groups	Percent that did not start as environ- mental group	Percent starting as environmental group
1845–1959	35	88.6	11.4
1960–1969	22	95.5	4.5
1970–1979	49	83.7	16.3
1980–1989	159	53.5	46.5
1990–1994	65	36.9	63.1

Compiled from the *People of Color Environmental Groups Directory* (Bullard, 1992b, 1994) and telephone interviews with environmental groups.

formed before 1960 started out as environmental groups, less than 5 percent of the groups that formed during the 1960s started out as environmental groups. This supports the argument that the political activism of the sixties around civil rights occupied the time and attention of many politically active people of color. During the 1970s, 16 percent of the groups started as environmental groups; by the next decade over 47 percent of the groups started as environmental groups, and during the 1990s, an impressive 63 percent of the groups were founded as environmental groups.

SIZE AND GEOGRAPHIC FOCUS

Although a common perception of people of color environmental groups is that of small neighborhood groups working on toxics, these groups often have a broader geographic focus. Thirty-seven percent of the groups focus solely on local issues, 11 percent work statewide, 21 percent are regional groups, and the remaining 21 percent are either national or international groups, and 12 percent are tribal. Just over half the groups have one hundred or fewer members; 17 percent of the groups have no members, while 38 percent have between one and one hundred members. Twenty-four percent of the groups have between 101 and 1,000 members, and the remaining 21 percent have more than 1,000 members.

In general, predominantly white environmental groups tend to have larger memberships. Whereas 79 percent of people of color environmental groups have one thousand or fewer members, 40 percent of the predominantly white environmental groups fall in this category (only 9 percent of the predominantly white groups have one hundred or fewer members).[44] The greatest number of predominantly white environmental groups (30 percent) have between 1,001 and 5,000 members.

GENDER AND LEADERSHIP

The importance of women in environmental groups of color and in the environmental justice movement cannot be understated. In no other sector of the environmental movement (not even in the more progressive or radical sectors) can one find such high percentages of women (particularly women of color) occupying leadership positions[45] or being responsible for such a variety of activities. Fifty-one percent of the leaders listed in the *People of Color Environmental Groups Directory* (Bullard, 1992b, 1994) are women. They are founders, presidents, or chief contact person for the group. In twenty-five of forty-one states, women comprised 50 percent or more of the leaders listed. A similar analysis of Schafer's (1993) study shows that 59 percent of the environmental justice groups profiled were led by women, many of whom were women of color. Similarly, about 48 percent of the delegates attending the People of Color Environmental Leadership Summit were women of color.[46]

In comparison, women are less likely to have leadership positions in predominantly white environmental organizations. The *Conservation Directories* (National Wildlife Federation, 1993, 1994) and the *Gale Environmental Sourcebook*[47] (Hill and Picirelli, 1992) were used to tabulate the number of organizations listed, the number of leadership positions listed in each organization, and the number of women listed in leadership positions. While 88 of the organizations (6.3 percent) listed no male leaders, 603, or 43 percent, of the predominantly white groups did not list any female leaders. Women occupied 27 percent of the leadership positions listed, but even when women were listed as leaders, they were often in an auxiliary position in the organization, i.e. administrative secretary, treasurer. They had little chance of occupying the top spot—the presidency, chair, or executive directorship—in the organization. Only 17 percent of the groups listed a woman as president or chair. In only two jurisdictions—Puerto Rico and the Virgin Islands—did women occupy the presidency in 50 percent or more of the organizations studied.

Gender, Leadership, and the Geographic Focus of the Organization

In people of color environmental groups, females were far more likely to be listed as leaders of local groups; 65 percent of the groups with a local focus are led by women, so are 42 percent of the statewide groups, 48 percent of the regional groups, and 51 percent of the international groups. Women are least likely to have leadership positions in tribal groups—only 39 percent of these groups listed women as leaders.

Gender, Leadership, and Regional Variation

Women were also more likely to be listed as leaders of environmental groups of color in the Southwest and South than in any other region of the country; they were far less likely to be leaders in the Pacific Northwest than in other regions (Table 5). Fifty-six percent of the leaders in the Southwest were female, compared to 55 percent in the South, 48 percent in the Midwest, 46 percent in the Northeast, and 35 percent in the Pacific Northwest. Women in predominantly white groups were least likely to be leaders in the South and most likely to be listed as leaders in groups in the Northeast (Table 6). While women are listed as leaders for 29 percent of the leadership positions in the Northeast, they occupy only 21 percent of those in the South, 26 percent of those in the Midwest, and 27 percent of those in the Southwest and Pacific Northwest.

The Rise of Environmental Justice Activism

In the euphoria over the new environmental consciousness sweeping the country during the sixties and seventies, the growing political power of a re-energized

Table 5 Regional Variation in Female Leadership in People of Color Environmental Justice Groups

Region	Number of leaders	Percent female leaders
Nationwide	344	50.9
Northeast/Mid-Atlantic	67	46.3
South/Southeast	99	54.5
Midwest	44	47.7
Southwest/West	108	55.6
Pacific Northwest	26	34.6

Compiled from the *People of Color Environmental Groups Directory* (Bullard, 1992b, 1994) and telephone interviews with environmental groups.

Table 6 Regional Variation in Female Leadership in Predominantly White Environmental Groups

Region	Number of leaders	Percent female leaders
Nationwide	7235	26.5
Northeast/Mid-Atlantic	3477	28.8
South/Southeast	973	21.0
Midwest	1094	26.4
Southwest/West	918	26.9
Pacific Northwest	773	27.0

Compiled from the *Conservation Directory* (National Wildlife Federation, 1993, 1994) and the *Gale Environmental Sourcebook* (Hill and Picirelli, 1992).

environmental movement and the readily discernible environmental gains, most failed to notice that pollution cleanup did not necessarily reach inner-city communities of color.[48] For example, in Washington, D.C., while the Potomac River was cleaned up to enhance tourism and other forms of recreation in the nation's capital, the Anacostia River, which runs through one of the city's African American communities, was not cleaned up. This neglect of African American and other people of color communities has led to declining air and water quality standards, increased toxic exposure, increased health risks, and declining quality of life.[49]

As discussed above, the most dramatic change in levels of mobilization of people of color occurred during the 1980s with the emergence of the environmental justice movement,[50] a radical, multiracial, grassroots environmental and social justice movement. This rapidly growing sector of the environmental movement is made up of thousands of grassroots environmental groups nationwide.[51] These groups collaborate with each other on varying levels and coordinate among themselves. Environmental justice activists come from an assortment of racial and social class backgrounds: they are African American, Latino, Asian, Native American, and white activists. Such a sociodemographic profile is unique in the environmental movement (see Table 7). Prior to the emergence of the environmental justice movement, the members of all the other sectors of the environmental movement were primarily white and middle class. Even the most radical environmental groups lacked racial and class diversity among their leadership, staff, volunteers, and membership.[52]

The environmental justice movement grew out of the dissatisfaction that grassroots activists had with the mainstream environmental agenda and the increasing environmental risks faced by people of color and poor communities. The advent of the environmental justice movement marks a radical departure from the traditional, reformist ways of perceiving, defining, organizing around, fighting, and discussing environmental issues; it challenges some of the most fundamental tenets of environmentalism that have been around since the late 1800s. It questions some of the basic postulates, values, and themes underlying the kinds of environmentalism characteristic of organizations like the Sierra Club, the Audubon Society, the Appalachian Mountain Club, the Izaak Walton League, the Nature Conservancy, and the National Wildlife Federation that are primarily concerned with wildlife, mountaineering, forest preservation, hunting, and fishing. It also challenges the types of environmentalism that arose in the sixties and seventies. The orientation of groups formed in this period ranged from radical, direct-action-oriented groups like Greenpeace[53] and Earth First! which focused their attention primarily on whales, nuclear disarmament, and forest preservation, to legal, technocratic, and lobbyist-oriented groups like Natural Resources Defense Council and the Environmental Defense Fund that focused on environmental laws and policies. Although many of the groups formed during the sixties and seventies started out as grassroots groups critical of the reform agenda of the pre-1960s environmental organizations, most of these post-Carson-era (1960–1979) environmental organizations eventually adopted reform

Table 7 Racial Composition of the Membership of People of Color Environmental Justice Groups

Racial composition of the membership	Number of groups	Percent of groups
African American	77	23.3
Latino	37	11.2
Asian	2	0.6
Native People	88	26.6
Mixture of people of color groups	88	26.6
People of color and whites	39	11.8

Compiled from the *People of Color Environmental Groups Directory* (Bullard, 1992b, 1994) and telephone interviews with environmental groups.

agendas, lost their close ties to the grassroots, and moved away from radical political dialogues and strategies. Like their predecessors, these groups did not adopt a social or environmental justice agenda, and they too lacked racial and social class diversity. They did little to try to understand the environmental problems affecting communities of color.

Several factors (some of which have been discussed above) have contributed to the evolution and sustained growth of the environmental justice movement and to the growing participation of people of color in environmental issues (see Box 2).

Box 2   Factors Contributing to Environmental
Justice Activism

1. Discovery of and prolonged exposure to toxins in many communities[54]
2. Deliberate targeting of communities of color for hazardous waste facilities[55]
3. Research and evidence linking race and exposure to environmental hazards and to detrimental health outcomes[56]
4. Corporate neighbors poisoning and eliminating communities of color[57]
5. Inconsistent and discriminatory policies of the Environmental Protection Agency and the courts[58]
6. Increased communication and dissemination of studies
7. A response to nimbyism (not in my backyard)
8. Organizing around broad themes of justice and fairness
9. Redefining what issues are considered "environmental"
10. Building alliances with preexisting movements and community institutions
11. Linking local concerns with regional, national, and international concerns
12. Linking environmental rights with civil rights[59]
13. Political strategies
14. Recognizing the leadership potential of women

TARGETING MINORITY COMMUNITIES

It is beyond the scope of this chapter to discuss all the aforementioned factors in detail, so only a few of the major factors will be discussed. First, the discovery of and prolonged exposure to toxins in many communities and the deliberate targeting of communities of color for hazardous waste disposal and the placement of undesirable facilities has incensed many. For example, predominantly African

American and Latino communities of south and eastern Los Angeles were out-raged to discover that their community was chosen for a huge incinerator project dubbed LANCER. Suspecting that they were chosen because they were the "path of least resistance" (low income, poorly educated, predominantly minority, low voter turnout), African American women of Concerned Citizens of South Central Los Angeles and Latinas from Mothers of East Los Angeles joined forces to lead a coalition of groups opposed to the siting of LANCER.[60] The efforts of these and other environmental justice activists have been buoyed by widely disseminated research and evidence indicating that the racial composition of a community plays a role in siting decisions and that race is linked with exposure to environmental hazards and to detrimental health outcomes.

Another catalyst for environmental justice activism is the growing practice of corporations poisoning neighboring communities and in some cases buying out and boarding up contaminated homes and attempting to relocate communities elsewhere. This is the fate of Revielletown and Morrisonville—the oldest Afri-can American community in the country, a community formed by freed slaves.[61]

RESEARCH AND THE CLAIMS OF DISPROPORTIONATE IMPACTS

Research played a key role in the growth of the environmental justice movement. Where research did not exist to substantiate claims being made by grassroots environmental groups, people of color and other activists conducted research that validated the claims. Most research points to disproportionate amounts of incinerators, landfills, and other noxious and hazardous facilities being sited in communities of color. In addition, a review of the medical, air pollution and other environmental health studies conducted over the past four decades also provide evidence to support these claims. Ongoing reviews of the literature on lead and other heavy metals also confirm that people of color are disproportion-ately exposed to and poisoned by some of these metals. Lead is one example where about 50 percent of all the children with elevated lead levels (ten micro-grams/deciliter) are African American.[62] People of color have the most danger-ous jobs and suffer the highest rates of job-related injuries.[63] For some occupational groups like farm workers, the Occupational Health and Safety Act either does not protect the workers (who are overwhelmingly people of color) at all or protects them inadequately.[64] Since the 1960s, air pollution studies have shown that people of color live in areas with very foul air. Although the Clean Air Act had the effect of making the air in the suburbs cleaner, the same was not always true for the inner city.[65]

EPA PROCEDURES AND POLICIES

Lavelle and Coyle's (1992) study "Unequal Protection: The Racial Divide in Environmental Law" is one of the most extensive analyses of the discriminatory impacts of environmental laws and policies. The study examined how the actions of the EPA (particularly the Superfund process) discriminate against people of color. Lavelle and Coyle found the following.

- Contaminated sites in many urban areas where people of color live were less likely to be recognized as Superfund sites than sites in other communities. Although communities like the predominantly African American Altgeld Gardens (the "toxic donut," as residents call it) have fifty toxic sites in a six-by-six-mile area, the Hazard Ranking System[66] (HRS) does not take into account the combined effect that these sites have on the ten thousand residents. The sites are evaluated and ranked separately. Despite criticisms, the EPA has been reluctant to modify the HRS.
- The HRS is biased in such a way that sites in rural areas are more likely to be placed on the National Priorities List (NPL) than other sites. Rural sites where residents drink well water are more likely to be placed on the NPL than urban sites connected to municipal water supplies. In the case of Altgeld Gardens, activist Hazel Johnson and her daughters Cheryl and Valerie formed People for Community Recovery—the only environmental (justice) group in this country located in a housing project. After they successfully campaigned to get municipal water hookup, they found that none of the fifty sites surrounding them were ranked highly enough to be placed on the NPL because the sites no longer threatened their drinking water supplies—the wells they drank from for years.
- It takes 20 percent longer to evaluate sites in minority communities to determine if they should be placed on the NPL than it does to evaluate sites in white communities.
- Once sites are placed on the NPL, it takes longer to clean up sites in communities of color than in white communities—an average of about 9.5 years in white communities and 13.1 years in communities of color.
- There is a difference in the type of cleanup options chosen for communities of color and white communities. While 45 percent of the sites in white communities were contained and 55 percent treated, 52 percent of the sites in communities of color were contained and only 48 percent treated.[67]
- The EPA, under Title VI, has a responsibility to ensure nondiscrimination in the use of pollution control funds, but early in the history of the agency, a decision was made not to emphasize the agency's civil rights enforcement responsibilities.[68]
- There were significant differences between the fines levied against companies violating the Resource Conservation and Recovery Act in communities of color and in white communities. Penalties averaged $55,318 in communities of color and $335,556 in white communities. On average, it is about six times more expensive to violate waste laws in white communities than in communities of color.
- If the decision is made to relocate communities, it takes longer to relocate communities of color, and residents of these communities are paid less

than residents in white communities in similar circumstances. For example, two communities (Mountain View Mobile Home park in Globe, Arizona, a white blue-collar community, and Carver Terrace, a middle-class African American community in Texarkana, Texas) were built about the same time on contaminated soil. Two years after the Mountain View residents started to wage a legal and political battle for justice, they each won a settlement for $80 thousand. It took the Carver Terrace residents twenty years to secure $30–40 thousand for their homes. Carver Terrace residents started their struggles soon after they moved into the development in 1968. Like Love Canal, whenever it rained, chemicals bubbled to the surface, strong odors fouled the air, and skin rashes and nose bleeds bothered residents. However, unlike Love Canal (a white community, in which residents were relocated and compensated for their homes soon after it was discovered that the site was contaminated), Carver Terrace residents were not relocated in a timely fashion.

THE COURTS

Even when the evidence clearly points to discrimination, people of color have a difficult time proving in court that they are being discriminated against. For instance, in cases involving the siting of facilities, it is not enough to prove that the *pattern* of siting is discriminatory and that the pattern results in disparate risks, one has to prove that the responsible party *intended to discriminate at the time of establishing the facility.* For example,

- In Houston (in 1979) where six out of the eight incinerators in the city were placed in African American communities, African Americans were unable to win their case.[69] The judge argued that they failed to prove discriminatory *intent* in the granting of the permit.
- Similarly, in 1991 in Virginia a district court judge argued that the selection of a site for a landfill in the predominantly African American King and Queen County did not violate the equal-protection clause.[70] This decision was reached despite the acknowledgment that over the preceding twenty years all the other landfills in the county were placed in areas that ranged from 95 percent to 100 percent African American.
- A federal judge in North Carolina rejected a challenge to the siting of the Warren County PCB landfill, which was to be located in the county with the highest percentage of African American residents in the state. The site did not meet EPA guidelines for the placement of landfills.[71] In Warren County, the EPA gave permission to build a landfill[72] to store PCB-laced soil in an area where the soil was highly permeable with only small amounts of clay present and where the water table was only seven feet below the bottom of the landfill. EPA guidelines stipulate that a landfill must be at least fifty feet above the groundwater and that such sites must be located where there are thick, relatively impermeable formations such as large-area clay pans.[73]

These are some of the research, policies, and practices that comprise part of the mounting evidence that makes the claim of environmental racism difficult to dismiss.[74] This evidence also makes it difficult for the mainstream environmental movement to claim that the benefits of its activities reach all citizens alike. Some in the environmental movement might feel that a discourse of environmentalism that includes variables like race, gender, and class is an irrelevant, political discourse concocted to make white, middle-class environmentalists feel uncomfortable and to divert attention from the important task of saving whales, trees, cute furry creatures, wilderness, and endangered species, and understanding global warming and reducing acid rain; as far as people of color and other environmental justice activists are concerned, no important environmental task can be accomplished (properly) unless the problem is embedded in a cultural, social, economic, and political context or reality.

FRAMING A NEW ENVIRONMENTAL DISCOURSE

The environmental justice movement is organized around principles that were articulated and formalized on October 27, 1991, by delegates attending the First National People of Color Environmental Leadership Summit. Delegates agreed on and adopted seventeen guiding principles that have since been widely accepted throughout the environmental justice movement (see Box 3). Given the evolutionary roots of the environmental groups of color—the movements, the institutions, and the social conditions out of which they emerge—it is not surprising that the principles espouse a strong proactive environmental stance combined with an identification of past injustice and an articulation for the need for future justice. The principles also underscore the need to expand the definition of what is considered "environmental." Environmentalists cannot continue to perceive, define, diagnose, and remediate environmental problems as if they are in a sphere completely divorced from any socioeconomic, cultural, and political reality. This expanded dialogue is a call for environmentalists to look at socioeconomic, cultural, and political factors in addition to the technological factors that they now rely on heavily.

People of color have pointed to these narrow, inflexible definitions as types of *discourses* that have an exclusionary or marginalizing effect on people who do not share the same perceptions, experiences, and worldview as those from the most dominant and powerful environmental groups, i.e., those who control the discourse. There is strong resistance to ideas, definitions, and discourses that originate from outside these organizations.[75] Because people of color groups define the environment more broadly; look at disproportionate effects of hazards on race, gender, age, and social class; and take new, innovative, and alternative approaches to solving environmental problems, they are often viewed with skepticism and distrust.

A comparison of surveys showing what environmental groups spend their resources on, and what issues they focus on, gives an indication of the scope

Box 3 Principles of Environmental Justice

PREAMBLE

WE, THE PEOPLE OF COLOR, gathered together at this multinational People of Color Environmental Leadership Summit, to begin to build a national and international movement of all peoples of color to fight the destruction and taking of our lands and communities, do hereby re-establish our spiritual interdependence to the sacredness of our Mother Earth; to respect and celebrate each of our cultures, languages and beliefs about the natural world and our roles in healing ourselves; to insure environmental justice; to promote economic alternatives which would contribute to the development of environmentally safe livelihoods; and to secure our political, economic and cultural liberation that has been denied for over 500 years of colonization and oppression, resulting in the poisoning of our communities and land and the genocide of our peoples, do affirm and adopt these Principles of Environmental Justice:

1. **Environmental justice** *affirms the sacredness of Mother Earth, ecological unity and the interdependence of all species, and the right to be free from ecological destruction.*
2. **Environmental justice** *demands that public policy be based on mutual respect and justice for all peoples, free from any form of discrimination or bias.*
3. **Environmental justice** *mandates the right to ethical, balanced and responsible uses of land and renewable resources in the interest of a sustainable planet for humans and other living things.*
4. **Environmental justice** *calls for universal protection from nuclear testing and the extraction, production and disposal of toxic/hazardous wastes and poisons that threaten the fundamental right to clean air, land, water, and food.*
5. **Environmental justice** *affirms the fundamental right to political, economic, cultural and environmental self-determination of all peoples.*
6. **Environmental justice** *demands the cessation of the production of all toxins, hazardous wastes, and radioactive materials, and that all past and current producers be held strictly accountable to the people for detoxification and the containment at the point of production.*
7. **Environmental justice** *demands the right to participate as equal partners at every level of decision-making including needs assessment, planning, implementation, enforcement and evaluation.*
8. **Environmental justice** *affirms the right of all workers to a safe and healthy work environment, without being forced to choose between an unsafe livelihood and unemployment. It also affirms the right of those who work at home to be free from environmental hazards.*
9. **Environmental justice** *protects the right of victims of environmental injustice to receive full compensation and reparations for damages as well as quality health care.*
10. **Environmental justice** *considers governmental acts of environmental injustice a violation of international law, the Universal Declaration On Human Rights, and the United Nations Convention on Genocide.*
11. **Environmental justice** *must recognize a special legal and natural relationship of Native Peoples of the U.S. government through treaties, agreements, compacts, and covenants affirming sovereignty and self-determination.*
12. **Environmental justice** *affirms the need for urban and rural ecological policies to clean up and rebuild our cities and rural areas in balance with nature, honoring the cultural integrity of all our communities, and providing fair access for all to the full range of resources.*
13. **Environmental justice** *calls for the strict enforcement of principles of informed consent, and a halt to the testing of experimental reproductive and medical procedures and vaccinations on people of color.*
14. **Environmental justice** *opposes the destructive operations of multi-national corporations.*
15. **Environmental justice** *opposes military occupation, repression and exploitation of lands, peoples and cultures, and other life forms.*
16. **Environmental justice** *calls for the education of present and future generations which emphasizes social and environmental issues, based on our experience and an appreciation of diverse cultural perspectives.*
17. **Environmental justice** *requires that we, as individuals, make personal and consumer choices to consume as little of Mother Earth's resources and to produce as little waste as possible; and make the conscious decision to challenge and reprioritize our lifestyles to insure the health of the natural world for present and future generations.*

Source: People of Color Environmental Leadership Summit, Washington, D.C., October 1991.

of their agendas and whether the agenda has been broadened (i.e., are people of color environmental groups and the rest of the environmental justice movement having an effect on the mainstream or more established environmental groups?). A 1988 survey of 248 staff members of predominantly white, reformist environmental organizations showed that environmental groups spent most of their resources on fish and wildlife protection, forest and park management, land stewardship, and other closely related issues. Although some groups spent money on toxics and hazardous waste management, these organizations really dealt with a narrow range of issues.[76]

The results of a 1992 survey conducted by the Environmental Careers Organization tells a different story.[77] It shows that the sixty-one reformist organizations surveyed have expanded their agenda somewhat but they still spend most of their time on wildlife, forest management, conservation, environmental education, fishing, water quality, and pollution. There was some overlap between reformist and environmental justice groups in the issues they worked on, but there were some distinct differences. Reformist organizations were far less likely to work on issues like public health, occupational health and safety, hazardous wastes, incineration, human rights/civil rights, and environmental justice than the seventy-six environmental justice groups surveyed. None of the reformist groups worked on issues like housing, neighborhood garbage/litter, Native land/water rights, farm worker/labor organizing, or military toxics. On the other hand, very low percentages of environmental justice groups worked on issues like conservation and depletion of the ozone layer.

Analysis of information presented in the *People of Color Environmental Group Directory* (Bullard, 1992b, 1994) enhances our understanding of what people of color environmental justice groups work on. Over 70 percent of them work on water pollution and toxics, while between 35 and 68 percent work on waste disposal, community organizing, environmental justice, air pollution, recycling, worker health and safety, housing, pesticides, energy, parks and recreation, and wildlife issues (Table 8). One in four people of color environmental justice groups are working on issues relating to the siting of facilities in their communities.

CONCLUSION

Environmental groups of color are, for the most part, radical conflict groups using an injustice frame to articulate their ideas. They recruit heavily from existing networks and use the existing dialogues and struggles in communities of color to amplify and expand the framing of the discourse. By linking the discourse about toxics with a discourse about racism, labor, and occupational health and safety, they have received support from a wide cross section of people. Such a response was not forthcoming (mobilization) when the discourse was structured primarily around forest and wildlife preservation.

Table 8 Percentage of people of color environmental groups focusing on selected issues

Issues	Number of groups	Percent
Water pollution	241	73.0
Toxics	239	72.4
Waste disposal	220	66.7
Community organizing	208	63.0
Environmental justice	194	58.8
Air pollution	191	57.9
Recycling	157	47.6
Worker health and safety	145	43.9
Housing	142	43.0
Pesticides	140	42.4
Energy	135	40.9
Parks and recreation	131	39.7
Wildlife	114	34.5
Lead poisoning	113	34.2
Facility siting	86	26.1
Asbestos	43	13.0

Compiled from the *People of Color Environmental Groups Directory* (Bullard, 1992b, 1994) and telephone interviews with environmental groups.

Scholars did not predict the current level of participation in environmental issues in communities of color or poor white communities. In studies of participation by African American and other people of color in environmental issues or in concern for the environment, scholars focused much of their attention on explaining nonparticipation or depressed levels of concern and action. Heavy emphasis was placed on Maslow's (1954) hierarchy of needs theory.[78] Researchers relying on this explanation failed to realize that a person who does not have all their basic needs met could still act on a seventh-order concern like the environment and elevate it to the highest priority if that issue was framed in life-or-death terms or in a manner that made it an immediate concern of the individual. This has happened in the case of toxics, where people have elevated the toxic threat to a level where they become politically active because they feel their lives depend on the effectiveness of their political activism.

People of color environmental groups have been very effective at framing the issues so that ordinary people (many of whom were never involved in political action before) have been able to see the relevance and the importance of the issue to their lives. These groups have supplemented the limited resources of the groups by establishing links with other community organizations; building alliances with regional, national, and international groups; and making connections with other movements like the civil rights movement, the labor movement, and religious institutions. This network of local and regional groups has exerted tremendous pressure on organizations such as the EPA and the most powerful environmental organizations. The pressure from people of color environmental justice groups has resulted in the EPA evaluating its programs and policies for evidence of disproportionate negative impacts on communities of color and establishing mechanisms for environmental justice initiatives within the agency and in communities across the country. Similarly, some major environmental groups like Greenpeace have made tremendous progress in collaborating with communities of color on environmental projects, and others like the National Wildlife Federation have established environmental programs for minority inner-city youth. A final example comes from the Environmental Careers Organization that has placed hundreds of minority students in environmental internships across the country.

Environmental justice activists, mainstream environmentalists, and environmental policy makers have embarked on a long and difficult dialogue. Despite the awkwardness, unease, and mistrust, these groups have to explore ways of bridging their differences and discovering their commonalities because the larger goal of living in a cleaner, safer environment binds them in a common struggle.

ENDNOTES

1. For the purposes of this chapter, the terms ''women of color'' and ''people of color'' will refer specifically to African Americans, Latinos/as, Native Americans, Asian Americans, and other ethnic minority groups in the United States.

2. Mitchell (1980, 44–45; 1979, 16–55), Kellert (1984, 209–228), Taylor (1989, 175–205; 1992), Hershey and Hill (1978, 339–358), Lowe, Pinhey, and Grimes (1980, 423–445), Buttel and Flinn (1978, 445), Van Liere and Dunlap (1980, 191–197), Buttel (1987, 465–488).

3. Hays (1959), Rosenbaum (1991).

4. Carson (1962).

5. The four eras of environmental mobilization are the pre-movement era, 1845–1913; the post-Hetch Hetchy era, 1914–1959; the post-Carson era, 1960–1979; and the post-TMI/Love Canal era, 1980–present.

6. The Hetch Hetchy controversy, which raged from 1910 to 1913, arose when the city of San Francisco proposed a dam on the Toulumne River (Hetch Hetchy) Valley in Yosemite. According to some, Hetch Hetchy was as spectacular as the Yosemite Valley, which was already designated a national park. The opponents of the Hetch Hetchy dam, John Muir and the Sierra Club, claimed the land should be *preserved* and not used in a utilitarian manner. It should be left untouched for future generations. The proponents of the dam, the City of San Francisco, Gifford Pinchot (the founder of the U.S. Forest Service), and many western congresspersons argued that Muir and his colleagues were being selfish. They argued that the land should be *conserved* or used wisely or sustainably. They argued that the good of the many (providing water for a parched city) should prevail over the good of a few (saving land for itself and for future generations). The controversy therefore pitted the preservationists against the conservationists (Nash, 1982).

7. Bramwell (1989), Fleming (1972), Fox (1985), Nash (1982), Paehlke (1989), Pepper (1986), Taylor (1992).

8. Snow and Benford (1992, 133–155).

9. National Wildlife Federation (1993), Hill and Piccirelli (1992). Although the *Conservation Directory* is not a complete listing of all environmental groups (it lists very few environmental justice groups—white or people of color groups, and only lists groups that have been in existence for at least three years), it is still one of the most comprehensive directories of the environment. The *Gale Environmental Sourcebook,* which lists more local, working-class, and environmental justice groups, was used to supplement the listing of the *Conservation Directory.*

10. All these movements were either in existence or trying to establish themselves at the turn of the century (see Fox, 1985; Kelly, 1996).

11. This paper uses the number of groups and the time period in which they were formed as an indicator of level of mobilization around environmental issues. Though there are problems with this indicator (the number of organizations formed does not give an indication of the size of the membership and the level or effectiveness of recruitment), given the limited data on formation and growth of environmental groups, it is one of the more trustworthy indicators of mobilization. Although many environmental groups report the size of their membership, not all do, and when they do, some inflate the numbers. In many cases there is a problem of double- and triple-counting; national groups sometimes report the membership of their local and state affiliates and these state and local affiliates either report local or state figures or the national figures. By relying heavily on the information in the *Conservation Directory* and the *Gale Environmental Sourcebook,* this analysis is biased toward groups that have been in existence for at least three years (minimum number of years they have to be in existence before they can be included in the *Conservation Directory*). Since this is not an attempt to analyze a census of environmental groups (founding dates were known for only 1,018 groups), the groups included in both directories provide a representative sample of environmental groups.

12. Bramwell (1989), Fleming (1972), Fox (1985), Nash (1982), Paehlke (1989), Pepper (1986), Taylor (1992).

13. Fox (1985).

14. Schwartz and Paul (1992, 205–223).

15. Schwartz and Paul (1992, 205–223), McCarthy and Wolfson (1992, 273–297).

16. Schwartz and Paul (1992, 205–223), McCarthy and Wolfson (1992, 273–297), McCarthy and Zald (1977, 1212–1241; 1973), McCarthy, Britt, and Wolfson (1991, 45–76).

17. Schwartz and Paul (1992, 205–223).

18. Carson (1962).

19. Taylor (1992, 28–54, 224–230), Environmental Careers Organization (1992), Mitchell (1980, 345–358).

20. Three Mile Island is a nuclear energy facility.

21. See Zald (1992, 326–348), McCarthy and Zald (1977, 1212–1241; 1973), Tarrow (1992, 174–202), Melucci (1989), Goffman (1974), Snow and Benford (1992, 133–155; 1988, 197–213), Foucault (1980, 1969, 1966), Klandermans (1992, 77–102).

22. Devall (1970, 123–126), Mohai (1991, 1985), Taylor (1989, 175–205), Harry, Gale, and Hendee, 1969, 246–254), Buttel and Flinn (1974), Hendee et al. (1968), Faich and Gale (1971, 270–287), Morrison, Hornback, and Warner (1972, 259–279), Lowe et al. (1980, 423–445).

23. Citizens' Clearinghouse for Hazardous Waste (1991, 2); Collette (1987, 44–45). Efforts to get independent confirmation of these estimates have proven futile. CCHW does not release or discuss its mailing list in any detail.

24. Mazmanian and Morell (1992, 27–28).

25. Downs (1972, 38–50).

26. For more on state legitimacy crises, see Habermas (1975).

27. Compiled from the *People of Color Environmental Groups Directory* (Bullard, 1992b, 1994) and from telephone interviews with environmental justice groups.

28. The *People of Color Environmental Groups Directory* (Bullard, 1992b, 1994) is a listing of environmental groups of color in the United States. Although this is not a census of environmental groups of color, it is an extensive listing of the groups existing in 1994.

29. Lavelle and Coyle (1992, S3).

30. Lavelle and Coyle (1992, S3).

31. W. Paul Robinson (1992, 153–162, 244).

32. Agency for Toxic Substances and Disease Registry (1993, 1–8).

33. See video, *We All Live Downstream* (Greenpeace, 1992).

34. Moses (1993, 161–178), videos *Wrath of Grapes* (United Farm Workers, 1986) and *No grapes* (United Farm Workers, 1993).

35. Santa Clara Center of Occupational Safety and Health (1993).

36. Intel Paper (1993).

37. Urban Environmental Conference, Inc. (1985b), Kreiss et al. (1981).

38. Lavelle and Coyle (1992), U.S. General Accounting Office (1983), United Church of Christ (1991c).

39. Urban Environmental Conference, Inc. (1985b), Urban Environment Conference, Inc. (1985a).

40. See Bryant and Mohai (1992).

41. Devall (1970, 123–126), Mohai (1991, 1985), Taylor (1989, 175–205), Faich and Gale (1971, 270–287), Mitchell (1980, 44–45; 1979, 16–55), Hershey and Hill (1978, 339–358), Lowe and Pinhey (1982, 114–128), Lowe, Pinhey, and Grimes (1980, 423–445), Buttel (1987, 465–488).

42. Friedman and McAdam (1992, 156–173).

43. A similar pattern was found in the study of minority environmental groups in Britain (Taylor, 1993, 263–296).

44. Some groups simply did not include any information about their membership. To categorize these as groups having no members would be inaccurate because several organizations with large memberships did not include this information in their profiles.

45. Leadership positions are defined as positions specified in the organizational profiles like president, chair, executive director, vice president, secretary, treasurer, program officer, program director, chief contact person. In the case of people of color environmental groups, all except one group listed only one leadership position. However, in the case of predominantly white groups, many leadership positions were listed. A hierarchy of the most important positions and the names of the persons occupying them were listed for all organizations (in the *Conservation Directory*, National Wildlife Federation (1993)).

46. United Church of Christ (1991a).

47. This directory contains many local or grassroots groups with women listed as leaders of the groups.

48. Gelobter (1992), Mann (1991).

49. Mann (1991), Commission for Racial Justice (1987), Bullard (1990), Environmental Protection Agency (1992), *Environmental Action Magazine* (1990).

50. Although there are whites involved in the environmental justice movement, this chapter focuses on the efforts of people of color in the United States to develop a movement that reflects the perspective of people of color.

51. Citizens' Clearinghouse for Hazardous Wastes (1991, 2), Collette (1987, 44–45), Bullard (1992a, 1991).

52. Paehlke (1989, 14–22), Fox (1985, 103–147), Devall (1970, 123–126), Mohai (1991, 1985), Taylor (1989, 175–205), Harry, Gale, and Hendee (1969, 246–254), Buttel and Flinn (1974), Hendee et al. (1968), Faich and Gale (1971, 270–287), Morrison, Hornback, and Warner (1972, 259–279), Lowe, Pinhey, and Grimes (1980, 423–445).

53. It should be noted that Greenpeace has been more responsive to the criticisms of environmental justice groups and people of color than any other major environmental organizations or any of the radical groups or sectors like Earth First!, greens, bioregionalists, ecofeminists, deep ecologists, or social ecologists. Although they continue with their traditional agenda, they have done extensive work on toxics and hazardous wastes. They have worked extensively with people of color communities and have demonstrated that they understand that in working with people of color communities, they need to be partners rather than superiors.

54. For detailed information on these and other sites, see Bryant and Mohai (1992), Bullard (1990, 1993), Hofrichter (1993), Schafer et al. (1993), and Mazmanian and Morell (1992).

55. Cerrel Associates, Inc. (1984), Trimble (1988), Blumberg and Gottlieb (1989).

56. Commission for Racial Justice (1987), Bullard (1993, 1990), Mann (1991), Environmental Protection Agency (1992), Bryant and Mohai (1992), Schafer et al. (1993).

57. Warden and Hirsch (1990), Lavelle and Coyle (1992).

58. Lavelle and Coyle (1992), *Bean v Southwestern Waste Management Corp.* (1979), and *R.I.S.E. v Kay* (1991).

59. Carson argued that individuals had a right to be protected from poisons applied by others into the environment and they should have a right to legal redress when that right is violated. Environmental justice groups are making a similar argument about harm and redress; groups argue that individuals have a right to a safe job, housing, and environment and that environmental rights cannot be separated from civil rights or environmental justice.

60. Blumberg and Gottlieb (1989, 58–84, 155–188), Cerrell Associates (1984), Trimble (1988).

61. Warden and Hirsch (1990), Lavelle and Coyle (1992), O'Byrne (1990), Greenpeace (1992).

62. Agency for Toxic Substances and Disease Registry (1988), Phoenix (1993).

63. Robinson (1991).

64. Farm workers are also the only workers not covered by minimum wage and child labor laws (Fair Labor Standards Act) or by laws giving them the right to organize and to collective bargaining (National Labor Relations Act). See Moses (1993, 161–178). For further discussion of Latino workers and inadequate protection under the occupational health and safety laws, see Moses (1993).

65. Gelobter (1992), Asch and Seneca (1978), Handy (1977), Berry (1977), Kurvant (1976), Freeman (1972).

66. In deciding whether a site should be designated a Superfund site, contaminated areas are evaluated to see if they are eligible to be placed on the National Priorities List (NPL). The EPA has developed a numerical rating scheme known as the Hazard Ranking System (HRS) to determine eligibility. The HRS considers seven factors: (1) the relative hazard to public health or the environment, taking into account the population at risk; (2) the hazardous potential of the substances at the site; (3) the potential for contamination of drinking water supplies; (4) direct contact with or destruction of sensitive ecosystems; (5) damage to natural resources that may affect the human food chain; (6) ambient air pollution; and (7) preparedness of the state involved to assume its share (typically, 10 percent) of the total costs and responsibilities for the cleanup. Sites receiving 28.5 or more on a scale of 1–100 are placed on the NPL. Sites on the NPL are eligible for Superfund monies (Mazmanian and Morell, 1992, 31; Wolf, 1988, 237).

67. Containment refers to the practice of dealing with the contamination on site; removal to another site or decontamination area is not an option. The toxics can be dug up, the area lined or relined against leaching, and the wastes buried on site; toxics can be incinerated on site or they can be neutralized on separated on site. Treatment refers to the process by which wastes are removed from the site and detoxified, buried, incinerated, etc., elsewhere. Of course, most communities dealing with toxic contamination prefer treatment procedures to containment procedures.

68. Under mounting criticisms from environmental justice organizations and increasing publicity generated by Lavelle and Coyle's (1992) study, the EPA has decided to look into its civil rights mission and take a case to court (Cushman, 1993).

69. *Bean v Southwestern Waste Management Corp.*

70. *R.I.S.E. v Kay.*

71. *NAACP v Gorsuch.*

72. PCBs were dumped illegally on back roads covering a large area of North Carolina. After considering several options, a decision was made to remove the contaminated soil and place it in a landfill, which was to be built in Warren County, the county with the largest African American population in the state.

73. LaBalme (1988, 23–30).

74. See Tesh and Williams (1994), Anderson et al. (1994), 83–100), Been (1994).

75. Foucault (1969, 1966).

76. Snow (1992, 55).

77. Environmental Careers Organization (1992, 44–45, 71–72). The comparisons were between reformist groups and social and economic justice (environmental justice) groups. The latter category had a mixture of people of color and predominantly white grassroots groups.

78. This theory argues that there are seven orders of needs that humans try to fulfill. The most basic needs, i.e., the desire for food, clothing, shelter, are filled first, and they preoccupy the individual until they are filled. The seventh-order needs, aesthetic needs (of which environmental concern is one), are of the highest order and are filled last. Therefore individuals concerned with gratifying their basic needs will not be concerned with aesthetic (environmental) needs. Academics studying environmentalism argued that because many people of color were concerned with meeting basic needs, they were less concerned (than whites) with environmental needs. Whites having met most of their basic needs were freer to concentrate on aesthetic needs.

LITERATURE CITED

Agency for Toxic Substances and Disease Registry. 1988. *The nature and extent of lead poisoning in children in the United States: A reprint to Congress.* Atlanta, Georgia: U.S. Department of Health and Human Services.

———. 1993. *Hazardous substances and public health.* Special topic: Native American/Alaska Native health concerns. Atlanta, Georgia: U.S. Department of Health and Human Services. February/March.

Anderson, Andy B., Douglas L. Anderton, and John Michael Oakes. 1994. Environmental equity: Evaluating TSDF siting over the past two decades. *Waste Age* (July).

Asch, P., and J. J. Seneca. 1978. Some evidence of the distribution of air quality. *Land Economics* 54:278–297.

Bean v Southwestern Waste Management Corp., 482 F. Supp. 673 (S.D. Tex. 1979).

Been, Vicki. 1994. Locally undesirable land uses in minority neighborhoods: Disproportionate siting or market dynamics? *Yale Law Journal* 102(6).

Berry, B. J. 1977. *The social burdens of environmental pollution.* Cambridge, England: Ballinger Publishing.

Blumberg, Louis, and Robert Gottlieb. 1989. *War on waste: Can America win its battle with garbage?* Covelo, CA: Island Press.

Bramwell, Anna. 1989. *Ecology in the 20th Century: A history.* New Haven, Connecticut: Yale University Press.

Bryant, Bunyan, and Paul Mohai, eds. 1992. *Race and the incidence of environmental hazards: A time for discourse.* Boulder, Colorado: Westview Press.

Bullard, Robert. 1990. *Dumping on Dixie: Race, class, and environmental quality.* Boulder, Colorado: Westview Press.

———. 1991. Environmental justice for all. *EnviroAction, Environmental News Digest for the National Wildlife Federation* (November).

———. 1992a. Environmental blackmail in minority communities. In *Race and the incidence of environmental hazards,* edited by Bunyan Bryant and Paul Mohai. Boulder, Colorado: Westview Press.

———, ed. 1992b. *People of color environmental groups directory 1992.* Flint, Michigan: The Mott Foundation.

———, ed. 1993. *Confronting environmental racism: Voices from the grassroots.* Boston, Massachusetts: South End Press.

———, ed. 1994. *People of color environmental groups directory 1994.* Flint, Michigan: The Mott Foundation.

Buttel, Frederick H. 1987. New directions in environmental sociology. *Annual Review of Sociology* 13:465–488.

Buttel, F. H., and W. L. Flinn. 1974. The structure and support for the environmental movement, 1968–70. *Rural Sociology* 39:56–69.

———. 1978. Social class and mass environmental beliefs: A reconsideration. *Environment and Behavior* 10:433–450.

Carson, Rachel. 1962. *Silent spring.* New York: Houghton Mifflin.

Cerrel Associates, Inc. 1984. Political difficulties facing waste-to-energy conversion plant siting. In *Waste-to-energy technical information series,* J. Stephen Powell, senior associate, chap. 3a. Los Angeles, California: California Waste Management Board.

Citizens' Clearinghouse for Hazardous Wastes. 1991. *Everyone's Backyard* 9(5): 2.

Collette, Will. 1987. Institutions: Citizens' clearinghouse for hazardous wastes. *Environment* 29(9).

Commission for Racial Justice. 1987. *Toxic waste and race in the United States: A national report on the racial and socioeconomic characteristics of communities with hazardous waste sites.* New York: United Church of Christ.

Cushman, John H. 1993. U.S. to weigh blacks' complaints about pollution. *New York Times,* November 11.

Devall, W. B. 1970. Conservation: An upper-middle class social movement. A Replication. *Journal of Leisure Research* 2(2): 123–126.

Downs, Anthony. 1972. Up and down with ecology—The issue-attention cycle. *Public Interest* 28.

Environmental Action Magazine. 1990. Special Issue. Beyond white environmentalism. 22(1).

Environmental Careers Organization. 1992. *Beyond the green: Redefining and diversifying the environmental movement.* Boston, Massachusetts: The Environmental Careers Organization.

Environmental Protection Agency. 1992. Environmental protection—Has it been fair? Special Issue. *EPA Journal* 18(1).

Faich, Ronald G., and Richard P. Gale. 1971. The environmental movement: From recreation to politics. *Pacific Sociological Review* 14(2): 270–287.

Fleming, D. 1972. Roots of the new conservation movement. *Perspectives in American History* 6.

Foucault, Michel. 1966. *The order of things: An archeology of the human sciences.* London: Tavistock.

———. 1969. *The archeology of knowledge.* London: Tavistock.

———. 1980. *Power/knowledge: Selected interviews and other writings, 1972–1977.* New York: Pantheon.

Fox, Stephen. 1985. *The American conservation movement: John Muir and his legacy.* Madison, Wisconsin: University of Wisconsin Press.

Freeman, A. M. 1972. The distribution of environmental quality. In *Environmental quality analysis,* edited by A. V. Kneese and R. M. Bower. Baltimore, Maryland: Resources for the Future.

Friedman, Debra, and Doug McAdam. 1992. Collective identity and activism: Networks, choices and the life of a social movement. In *Frontiers of social movement theory,* edited by Aldon M. Morris and Carol McClurg Mueller. New Haven, Connecticut: Yale University Press.

Gelobter, Michael. 1992. Toward a model of environmental discrimination: How environmental laws discriminate against low-income and minority communities. In *Race and the incidence of environmental hazards,* edited by Bunyan Bryant and Paul Mohai. Boulder, Colorado: Westview Press.

Gianessi, L. P., H. M. Peskin, and E. Wolff. 1977. The distributional impacts of national air pollution damage estimates. In *The distribution of economic well-being,* edited by F. T. Juster. Cambridge, England: Ballinger Publishing.

Goffman, Erving. 1974. *Frame analysis: An essay on the organization of experience.* New York: Harper.

Greenpeace. 1992. *We all live downstream.* Video. Washington, D.C.

Habermas, Jurgen. 1975. *Legitimation crisis,* Translated by Thomas McCarthy. Boston, Massachusetts: Beacon Press. (Original title: *Legitimationsprobleme im Spatkapitalismus.* Frankfurt: Suhrkamp, 1973.)

Handy, F. 1977. Income and air quality in Hamilton, Ontario. *Alternatives* 6(3): 18–24.

Harry, J., R. Gale, and J. Hendee. 1969. Conservation: An upper-middle class social movement. *Journal of Leisure Research* 1(2): 255–261.

Hays, S. 1959. *Conservation and the Gospel of efficiency: The progressive conservation movement.* Cambridge, Massachusetts: Harvard University Press.

Hendee, J. C., W. R. Catton, I. D. Marlow, and C. F. Brockman. 1968. Wilderness users in the Pacific Northwest—Their characteristics, values and management preferences. Research Paper PNW-61. Portland, Oregon: USDA Forest Service Forest Range Experiment Station.

Hershey, M. R., and D. B. Hill. 1978. Is pollution 'a white thing'? Racial differences in preadults' attitudes. *Public Opinion Quarterly* 41.

Hill, K., and A. Piccirelli. 1992. *Gale environmental sourcebook.* Detroit, Michigan: Gale Publishing.

Hofrichter, Richard, ed. 1993. *Toxic struggles: The theory and practice of environmental justice.* Philadelphia, Pennsylvania: New Society Publishers.

Intel Paper. 1993. Electronics Industry Good Neighbor Campaign.

Kellert, S. R. 1984. Urban American perceptions of animals and the natural environment. *Urban Ecology* 8:209–228.

Kelly, John. 1996. *Leisure.* Englewood Cliffs, New Jersey: Prentice–Hall.

Klandermans, Bert. 1992. The social construction of protest and the multiorganizational fields. In *Frontiers of social movement theory,* edited by Aldon Morris and Carol McClurg Mueller. New Haven, Connecticut: Yale University Press.

Kreiss, K., M. M. Zack, R. D. Kimbrough, et al. 1981. Cross-sectional study of a community with exceptional exposure to DDT. *Journal of the American Medical Association* 245:1926–1930.

Kurvant, W. J. 1976. People, energy, and pollution. *The American energy consumer,* edited by D. K. Newman and D. Day. Cambridge, Enlgand: Ballinger Publishing.

LaBalme, Jenny. 1988. Dumping on Warren County. In *Environmental politics: Lessons from the grassroots,* edited by Bob Hall. Durham, North Carolina: Institute for Southern Studies.

Lavelle, Marianne, and Marcia Coyle, eds. 1992. The racial divide in environmental law. Unequal protection. *National Law Journal,* Supplement, September 21, 1992.

Lowe, G. D., and T. K. Pinhey. 1982. Rural-urban differences in support for environmental protection. *Rural Sociology* 47.

Lowe, G. D., T. K. Pinhey, and M. D. Grimes. 1980. Public support for environmental protection: New evidence from national surveys. *Pacific Sociological Review,* 23.

Mann, Eric. 1991. *L.A.'s lethal air: New strategies for policy, organizing, and action.* Los Angeles, California: Labor/Community Watchdog Strategy Center.

Maslow, A. H. 1954. *Motivation and Personality,* 2d ed. New York: Viking Press.

Mazmanian, Daniel, and David Morell. 1992. *Beyond superfailure: America's toxics policy for the 1990s.* Boulder, Colorado: Westview Press.

McAdam, Doug. 1982. *Political processes and the development of black insurgency: 1930–1970.* Chicago: University of Chicago Press.

McCarthy, John D., David W. Britt, and Mark Wolfson. 1991. The institutional channeling of social movements in the modern state. *Research in Social Movements: Conflict and Change* 13:45–76.

McCarthy, John D., and Mark Wolfson. 1992. Consensus movements, conflict movements, and the cooptation of the civic and state infrastructures. In *Frontiers of Social Movement Theory,* edited by Aldon Morris and Carol McClurg Mueller. New Haven, Connecticut: Yale University Press.

McCarthy, John D., and Mayer N. Zald. 1973. *The trend of social movements in America: Professionalization and resource mobilization.* Morristown, New Jersey: General Learning Press.

———. 1977. Resource mobilization and social movements: A partial theory. *American Journal of Sociology* 82:1212–1241.

Melucci, Alberto. 1989. *Nomads of the present: Social movements and individual needs in contemporary society.* Philadelphia, Pennsylvania: Temple University Press.

Mitchell, Robert C. 1979. National environmental lobbies and the apparent illogic of collective action. In *Collective Decision Making,* edited by C. Russell. Baltimore, Maryland: Johns Hopkins University Press.

———. 1980. *Public opinion on environmental issues, results of a national public opinion survey.* CEQ, DOA, DOE, and the EPA. Washington, D.C.: Government Printing Office.

Mohai, Paul. 1985. Public concern and elite involvement in environmental conservation issues. *Social Science Quarterly* 55(4): 820–838.

———. 1991. Black environmentalism. *Social Science Quarterly* 71(4): 744–765.

Morrison, Denton E., K. E. Hornback, and W. K. Warner. 1972. The environmental movement: Some preliminary observations and predictions. In *Social behavior and natural resources and the environment,* edited by W. Burch Jr. et. al. New York: Harper and Row.

Moses, Marion. 1993. Farm workers and pesticides. In *Confronting environmental racism: Voices from the grassroots,* edited by Robert Bullard, Boston, Massachusetts: South End Press.

NAACP v Gorsuch, 82–768 (E.D.N.C. 1982).

Nash, Roderick. 1982. *Wilderness and the American mind,* 3rd. ed., New Haven, Connecticut: Yale University Press.

National Wildlife Federation. 1993. *Conservation directory,* 38th ed. Washington, D.C.: National Wildlife Federation.

———. 1994. *Conservation directory,* 39th ed. Washington, D.C.: National Wildlife Federation.

O'Byrne, James. 1990. The death of a town. *The Times Picayune,* February 20.

Paehlke, Robert. 1989. *Environmentalism and the future of progressive politics.* New Haven, Connecticut: Yale University Press.

Pepper, David. 1986. *The roots of modern environmentalism.* London: Croom & Helm, Ltd.

Perfecto, Ivette. 1992. Pesticide exposure of farm workers and the international connection. In *Race and the incidence of environmental hazards,* edited by B. Bryant and P. Mohai. Boulder, Colorado: Westview Press.

Phoenix, Janet. 1993. Getting the lead out of the community. In *Confronting environmental racism: Voices from the grassroots,* edited by Robert Bullard. Boston, Massachusetts: South End Press.

R.I.S.E. v Kay, 768 F. Supp. 1144 (E.D. Va. 1991).

Robinson, James. 1991. *Toil and toxics: Workplace struggles and political strategies for occupational health.* Berkeley, California: University of California Press.

Robinson, W. Paul. 1992. Uranium production and its effects on Navajo communities along the Rio Puerco in western New Mexico. In *Race and the incidence of environmental hazards: A time for discourse,* edited by B. Bryant and P. Mohai. Boulder, Colorado: Westview Press.

Rosenbaum, Walter. 1991. *Environmental politics and policy.* 2nd ed. Washington, DC: Congressional Quarterly Press.

Santa Clara Center for Occupational Safety and Health. 1993. *Working Healthy* (December): 1–12.

Schafer, Kristin, et al. 1993. *What works: Local solutions to toxic pollution.* Washington, D.C.: The Environmental Exchange.

Schwartz, Michael, and Shuva Paul. 1992. Resource mobilization versus the mobilization of people: Why consensus movements cannot be instruments of social change. In *Frontiers of Social Movement Theory,* edited by Aldon Morris and Carol McClurg Mueller. New Haven, Connecticut: Yale University Press.

Snow, Donald. 1992. *Inside the environmental movement: Meeting the leadership challenge.* Covelo, California: Island Press.

Snow, David A. and Robert D. Benford. 1988. Ideology, frame resonance, and participant mobilization. *International Social Movement Research* 1:197–217.

———. 1992. Master frames and cycles of protest. In *Frontiers of social movement theory,* edited by Aldon Morris and Carol McClurg Mueller. New Haven, Connecticut: Yale University Press.

Tarrow, Signey. 1992. Mentalities, political cultures, and collective frames: Constructing meanings through action. In *Frontiers of social movement theory,* edited by Aldon Morris and Carol McClurg Mueller. New Haven, Connecticut: Yale University Press.

Taylor, Dorceta E. 1989. Blacks and the environment: Toward an explanation of the concern and action gap between blacks and whites. *Environment and Behavior* 21(2): 175–205.

———. 1992. Can the environmental movement attract and maintain the support of minorities? In *Race and the incidence of environmental hazards,* edited by Bunyan Bryant and Paul Mohai. Boulder, Colorado: Westview Press.

Tesh, Sylvia, and Bruce A. Williams. 1994. Science, identity politics, and environmental racism. Paper presented at the American Political Science Association, New York, September.

Trimble, Lillie C. 1988. What do citizens want in siting of waste management facilities? *Risk Analysis* 8(3).

United Church of Christ. 1991a. The First National People of Color Environmental Leadership Summit summary report of delegate registrations, October 15, 1991. New York: United Church of Christ.

United Church of Christ. 1991b. *National people of color environmental leadership summit program guide.* New York: United Church of Christ.

United Church of Christ. 1991c. Proceedings of the First National People of Color Summit. Washington, D.C. New York: United Church of Christ.

United Farm Workers. 1986. *Wrath of grapes.* Video.

———. 1993. *No grapes.* Video.

U.S. General Accounting Office. 1983. *Siting of hazardous waste landfills and their correlation with racial and economic status of surrounding communities.* Washington, D.C.: U.S. Government Printing Office.

Urban Environmental Conference, Inc. 1985a. *Environmental cancer: Causes, victims and solutions.* Washington, D.C.: Urban Environmental Conference, Inc.

Urban Environmental Conference, Inc. 1985b. *Taking back our health: An institute on surviving the toxics threat to minority communities.* Washington, D.C.: Urban Environmental Conference. Inc.

Van Liere, K. D., and Riley Dunlap. 1980. The social bases of environmental concern: A review of hypothesis, explanations, and empirical evidence. *Public Opinion Quarterly* 44(2): 181–197.

Warden, A. C., and Karen Hirsch, producers. 1990. We all live downstream. Oakland, California: The Video Project.

Wolf, Sidney M. 1988. *Pollution law handbook: A guide to federal environmental laws.* Westport, Connecticut: Quorum Books.

Zald, Mayer N. 1992. Looking backward to look forward: Reflections on the past and future of the resource mobilization research program. In *Frontiers of social movement theory,* edited by Aldon Morris and Carol M. Mueller. New Haven, Connecticut: Yale University Press.

Part II

Adaptive
Visions—New
Perspectives,
Ecosystem
Management,
Leadership, and
Bioregionalism

Overview

This section of the book provides a critical examination of the evolution of some of the visions of natural resource agencies as they seek to order their response to some of the challenges outlined in Part I. The authors in this part contribute to our understanding of some of the underlying aspects of the ''new perspectives'' approach and how that led into the now favored ''ecosystem management'' approach. As these visions pass through the screens of external and internal interests, what was thought to be a better structure now becomes more of just another passing process.

Two other visions—leadership and bioregionalism—rise and fall with the tides of political influence and are included here, as they too reflect hopes of order but gain only more change. The popularity of courses, modules, and talk on leadership among people in the natural resource area has had a period of steady growth as the public has pressed for greater participation in the resource decisions that influence their lives, interests, and passions. One could, of course, point out that leaders should be so busy leading that they have little time to take or to teach courses on leadership, or that people who usually gain their positions by appointment from others, which usually means that they have exhibited the best conformity to the interests of the person doing the appointing, are not likely to do much actual leading. That is, high-ranking bureaucrats or executives are usually appointees rather than leaders who mobilize the masses.

Bioregionalism is a vision that has a fairly substantial history. The Tennessee Valley Authority was a very concrete response to the vision, but then none of the other regions—Columbia River, Connecticut River, and so forth—were organized. Indeed, the TVA has its enemies from both the right and the left of the political spectrum. Perhaps the scale was too grand, and the more modest watersheds that are well within a single political unit are the answer for the revitalization of the vision. Certainly the emergence of interest in ecosystem management and the rise of landscape ecology as a special kind of ecology will be strong influences upon a more chastened but more useful bioregional approach.

William R. Burch Jr.

Chapter 5

From New Perspectives to Ecosystem Management: A Social Science Perspective on Forest Management[1]

Roger N. Clark

George H. Stankey

Linda E. Kruger

There is nothing permanent except change.

Heraclitus (540–475 BC)

Today's world of natural resource use and management is fraught with conflict, confusion, and uncertainty. Public concern about forests in North America and around the globe is creating pressure for fundamental changes in forest management goals and the means to achieve them. Business as usual is no longer acceptable, but what constitutes an acceptable alternative is unclear and will undoubtedly be controversial.

Much of the information presented here has been drawn from papers, workshops, meetings, and field trips, along with a delphi survey in 1991 asking respondents about their views and evaluations of the New Perspectives program. In this chapter, we describe the background from which New Perspectives and ecosystem management emerged, the results of the 1991 New Perspectives Delphi, and alternative natural resource paradigms, and we outline some of the necessary conditions for integration across disciplines, functions, and organizations, a key goal of New Perspectives and, now, ecosystem management. We outline some of the implications of this discussion for management, education, and research and close with some "rules" about the relationships between forestry professionals and the society they serve.

WHY SOMETHING NEW?

The conditions leading to new programmatic initiatives, such as New Perspectives and ecosystem management, in the U.S. Forest Service are not unique but are common to most, if not all, natural resource institutions. Two factors have been significant in driving change. First, ecologists and other natural scientists have improved our understanding of the complexity of ecosystems and the cumulative effects of human activities on them. And second, society has become increasingly concerned about the values for which forests are being managed.

These factors have resulted in public and professional support for new approaches to forest management. This different approach was intended to recognize and accommodate the complexity of the nonhuman system while responding to public demands for sustainable resource management that provides a wide range of values and uses.

Recognition of the need for a new approach to natural resource management has come amidst conflict and controversy about the expectations society has for forests and how they are managed. Social values—the diverse values the public has for forests—encompass far more than commodity values such as timber and other material forest products. Foresters are good at providing traditional outputs, but perhaps less so at recognizing and managing for other forest values. A new approach needs to reflect a better understanding and incorporation of diverse societal values in planning and decision-making processes.

Traditionally, agencies such as the U.S. Forest Service have tried to serve the public interest and provide for the values the public desired from forests. What is it that characterized the situation in forestry such that a "new perspective" was needed? We suggest the following (Stankey and Clark, 1992):

> The public's diminished trust in the forestry profession has led to the feeling that forest management activities do not represent the broad public interest.
> There is a growing sense of dissatisfaction on the part of some segments of society with the direction that forestry programs have taken (i.e., an emphasis on the production of wood fiber versus other values such as wildlife and recreation) and with the processes through which allocation decisions are made.
> The increasing visibility of forestry due to the growing number of lawsuits, congressional action, and media coverage has politicized the decision-making environment within which natural resource management occurs.
> Fragmented administrative, organizational, and disciplinary structures and institutions adversely affect forestry. Reductionism artificially constrains the way in which problems are defined as well as the search for solutions.

Ecologists, social scientists, managers, and members of the public are concerned with the spatial and temporal dimensions of proposed programs as well as the linkages between the different components of the ecosystem. There is a need to identify the cumulative ecological and social effects of proposed programs in ways that allow people to better understand the long-term consequences of management alternatives.

Much of the public has shifted its attention from commodities and services provided by forests to environments, habitats, and ecosystem values. Moreover, the public is increasingly concerned with having an opportunity to participate in decisions about their forests.

These and other issues provided the impetus for development of New Perspectives and ecosystem management. In many respects, these driving issues constitute the very challenges forest management organizations must address to survive.

THE DOMINANT PARADIGM IN NATURAL RESOURCE MANAGEMENT

Some would argue that the New Perspectives program and ecosystem management represent a paradigm shift. A paradigm is a system of standards, behavioral norms, and conceptual approaches to problem solving (Kuhn, 1970). Why is the concept of paradigms important? The short answer is that the paradigm each of us accepts influences how we think and act (Stankey et al., 1992).

The dominant paradigm in resource management has been a rational, scientifically based, analytic process (Wondolleck, 1988). Within this paradigm, commodity values determined in the marketplace receive more attention than values not as easily identified, verified, and measured. As a result, the prevailing use of forest lands has been primarily timber production. Management activities have been viewed largely as discrete actions, and management has been functional in both organizational structure and funding. A boundary orientation has inhibited work across ownerships, disciplines, management units, and cultures.

AN EMERGING PARADIGM—WHAT'S NEW?

Elements of an alternative emerging paradigm call attention to a wide range of values, including many with no readily determined market value such as historical, cultural, wildlife, recreational, amenity, and spiritual values (Clark and Brown, 1990). Decision making in this alternative paradigm is open to the full community of interests and is truly participatory. A concept of sustainability of forest values and uses encompasses not only the biological and ecological, but social components as well. Improved communication between citizens and professionals results in information needed to make informed decisions with an understanding of the consequences and implications of decisions. Recognition and legitimization of knowledge held by the public provides the cornerstone in mutual learning between diverse publics and professionals.

Other characteristics of a new paradigm include integration of functions and disciplines within and across agencies and institutions; collaboration between all stakeholders, including nontraditional clients and interests; analysis at multiple temporal scales, including increased attention to long-term cumulative effects in both the natural ecosystem and social system; analysis at multiple spatial scales—microsites to landscapes to regional to global; and a legitimization of diverse values and uses.

MANAGING FORESTS FOR SOCIAL VALUES

The New Perspectives and ecosystem management programs emerged within a context of conflicting values about the management of forests. But the concept of values is elusive. Every social science discipline studies values and the way people attribute value to things. The literature on values provides an assortment of definitions, categories, groupings, and typologies. Although there is no one commonly accepted definition, most agree that "values are the standards that guide or determine attitudes and behavior" (Dunlap et al., 1983).

Values are measures of importance and worth to society. All values (including ecological values that are socially defined) are social values (Stankey and Clark, 1992). Some values are more easily recognized and measured, especially for goods and services that are exchanged in the marketplace and whose worth can be "measured" in dollars (Stankey et al., 1992). Stankey and Clark (1992) developed a classification system of six categories of social values related to forest resources: commodity values (timber, range, minerals), amenity values (scenery, wildlife, nature), environmental quality values (air and water quality), ecological values (habitat conservation, sustainability, biodiversity), public use values (subsistence, recreation, tourism), and spiritual values.[2] Koch (1990) describes social values as those values that benefit society as a whole or that a group of people desire, as distinct from individual and personal values.

RESPONDING TO DIVERSE VALUES: FOCUS ON PEOPLE, PLACES, AND PROCESSES

Three elements around which most social value problems and issues are identified are people (including their distribution, values, organization, and behavior), places (both geographic and symbolic dimensions at a variety of scales), and processes (the ecological processes and human activities and institutions that affect people, places, and their interactions). This "three p's" approach provides a framework within which to consider the issues and concerns that have given rise to the New Perspectives and ecosystem management programs (Stankey and Clark, 1992).

Although it is possible, and often essential, to describe each of the three components individually, it is more commonly the interaction between and among people, places, and processes that represents the most interesting and difficult issues. This suggests that integrated approaches for planning, management, and research will be necessary to protect, create, or enhance forest and other natural resources in keeping with societal values.

ALTERNATIVE PERSPECTIVES ON NEW PERSPECTIVES

In light of existing conditions and the increasing need for improved understanding of social values, how well have the alternative approaches to forest management represented by programs such as New Perspectives and ecosystem management responded? As an illustration, we turn to the results of the Delphi study we undertook in 1991, involving people from a variety of perspectives (managers, researchers, interest groups) and from across the United States. When we asked them to define what the idea of "New Perspectives" meant, six perspectives emerged (Clark and Stankey, 1991):

1. An ecologically founded approach to forest management: an approach that is more holistic and involves ecologically based information and principles.

2. The need for a greater integration of different forest uses and values: the approach will focus more attention on nontimber outputs and interactions among management activities and uses. It will provide for greater integration of a multitude of forest values and uses.

3. The need to focus on changing public values and uses of forest resources: this new way of doing business is an adaptive response to adjust to changing public values for forest resources.

4. The need for different approaches to making decisions about forest management: an approach that encompasses new processes for decision making that will lead to changes in forest uses and outputs and increased public participation.

5. Application of better management tools and improved knowledge of management consequences: the new approach involves an enhanced role for science and technology, reliance on technical solutions, and a rational-scientific approach to management.

6. Questioning agency and professional motives: not a new approach at all, merely a public relations ploy, propaganda, or a guise under which to continue business as usual.

Three themes underlie these perceptions. New Perspectives was seen as a philosophical concept, a set of specific practices and prescriptions, and as a new approach to doing business. As conceived by the U.S. Forest Service, New Perspectives did not prescribe a particular set of practices, but provided a framework within which to recognize the role of diverse values in natural resource decision making. Thus New Perspectives represented an evolutionary stage in an ongoing process in how the U.S. Forest Service and other land managers approach resource management.

PROBLEMS ASSOCIATED WITH INCORPORATING A SOCIAL COMPONENT IN FOREST MANAGEMENT

Stankey and Clark (1992) identify six problems confronting incorporation of a social component into any alternative forest management paradigm. However, by identifying these problems or barriers, we gain a structure within which to begin to increase knowledge and understanding of the relationship between social values and forest management; these barriers are likely to be equally applicable in other natural resource management areas, such as wildlife and fisheries:

Problem 1. Agencies have only a limited ability to integrate the various forms of social values—e.g., commodity, amenity, ecological, scientific, and spiritual—into the decision-making process.
Problem 2. There is an inadequate understanding of the values the public has for natural resources.
Problem 3. There is an inadequate understanding of what constitutes "acceptability" of forestry practices and the conditions resulting from the use of new techniques.
Problem 4. There are few effective mechanisms and processes available to translate public involvement into public participation in decision making.
Problem 5. The structure, processes, and values of natural resource organizations and professions limit the integration and incorporation of a full range of public values into decision making.
Problem 6. As currently structured, decision-making environments may be threatening, thus limiting the opportunity for debate and discussion of critical issues by resource managers, citizens, and others.

Several broad premises underlie the program of social values research to address these problems. First, a research, development, and applications philosophy must prevail in which the overriding concern is the delivery of solutions consistent with the needs of the ultimate client—the public. Second, continuing attention needs to be paid to integrating the physical-biological and social-economic aspects of the problems addressed above. Third, rapid changes in technology, knowledge, and public concerns and interests necessitate that the approach taken have an anticipatory/predictive dimension. And fourth, we must recognize that many of the problems faced today do not have unequivocal "answers" or solutions; we may have to be satisfied with more or less useful solutions.

INTEGRATION—EASY TO SAY, HARD TO DO

Implicit in New Perspectives and ecosystem management is the concept of integration. Problems one and five above specifically address integration, and others address it indirectly. Although integration is an important principle, there are many difficulties in actual implementation. We find little evidence of approaches that successfully integrate diverse social values into resource decision making. Most often, social values are treated as constraints.

In order to implement management that can integrate the variety of values the public demands, Clark and Brown (1990) suggest that the following fundamental conditions must be met.

A clear and comprehensive definition of what integrated resource management is and is not, and specific criteria for evaluating when integration has been achieved, are described.

Assumptions underlying integrated planning and management are clearly stated.

Managers and the public collaborate in the development of a range of alternative desired futures, which are then communicated by management so that people from diverse social and cultural backgrounds can understand where and when changes will affect them.

Unambiguous multiple value objectives clarifying the desired futures are specified.

Neighboring landowners and managers collaborate in defining desirable future conditions, their respective role in achieving them, and management prescriptions.

Professionals are open to new ways to manage for diverse values and to share power in decision making.

Enough interest groups and individuals seek the middle ground to reassert its value. Polarization is disruptive and will result in solutions that limit the range of options.

Aggressive monitoring and evaluation and a willingness and ability to adjust to new information characterizes management. Management is seen as an activity within a learning process.

Research helps professionals and the public learn what works and what is ineffective. It needs to focus on issues of social values, manager and public participation, landscape-level management, and integrated multiple-use management. To be most useful, research must be conducted across agencies at both the landscape and site-specific levels. Cross-disciplinary studies are needed to improve our understanding of the interrelationships between multiple resource values and uses. Synthesis and interpretation of knowledge from past research and management experience, coupled with demonstrations to test what does and does not work are needed.

Implications . . .

Sometimes we stare so long at a door that is closing
that we see too late the one that is open.

Alexander Graham Bell

Presuming that what we have offered in the preceding pages is valid, what does it mean for how we do business in the future? We suggest that there may be some changes in order for managers, educators, and researchers.

. . . for management

Natural resource professionals need to take the lead in interacting with members of the public to address complex problems. They need to share their ideas and listen to the diverse ideas held in society. Such interaction must be a two-way process within which mutual learning takes place.

Integration of multiple values and outputs requires more responsive administrative decision-making structures. Integrative policies, programs, and actions are needed. We cannot rely on the economic model to identify the importance of all values for which managers are responsible. The functional structures of management organizations inhibit creation or implementation of integrative processes. We need to acknowledge the importance of social values and find ways to integrate broader interests into decision making. Shared decision making is critical if people are going to be part of solutions rather than adding to or becoming the problem. As DePree (1989, 24–25) suggests,

> Everyone has the right and the duty to influence decision making and to understand the results. Participative management guarantees that decisions will not be arbitrary, secret, or closed to questioning.

. . . for education

Efforts are needed to break down disciplinary boundaries and encourage integrative approaches. Educational institutions need to refocus and become responsive to changing public perceptions and the diverse values of forests. A reevaluation of skills and substantive training is necessary to address current social and political, as well as biological and physical, conditions. Basic education needs to be broadened to promote critical thinking, problem solving, information management, and communication skills, in addition to traditional technical skills. Continuing education is imperative if professionals are to stay informed of changing conditions, knowledge and tools, and means of responding to changing societal expectations and priorities. Basic and continuing education should include information about the social-political context within which forestry is practiced.

. . . for research

New ways to think about research and how it can be more effective are needed. Innovative solutions are needed to identify organizational structures that encourage collaboration. Increased opportunities for researchers to work with managers and policy makers are needed to ensure that the knowledge being applied is the best available for the problem. Monitoring and adapting management approaches as new information becomes available are also necessary. Attention should be given to the integration of physical-biological, social-economic aspects of research studies. Although many problems do not have absolute right or wrong answers, improved understanding of consequences and implications of a full range of alternatives will enhance decision making at all levels.

ROLES AND RULES—SOME FINAL THOUGHTS

What lessons regarding how we go about incorporating social values in resource management have emerged from the experience gained to date in New Perspectives and ecosystem management? We suggest the following.

We must learn before we can teach. Whether we are scientists, managers, or concerned citizens, we need a better understanding of the social-political context within which we all operate, and an understanding of the perspectives of all stakeholders. In this regard, effective communication requires not only listening but understanding. Our values, biases, and the paradigms we each operate within filter what we hear from others and what they hear from us. Trained and learned incapacities resulting from education and experience are as deadly for scientists and managers as they are for the public. We may choose not to agree, but we must understand one another and value diverse points of view. The experience in New Perspectives, reinforced in our efforts to implement ecosystem management, is that our planning and management processes must foster open public participation and scrutiny and must aggressively seek opportunities for shared learning.

Learn the lessons, not just the vocabulary—walking the walk is critical. In the United States and abroad there is considerable public distrust of institutions, government, and professions. Skepticism and cynical views mean that actions will be evaluated, not slogans or labels. As we all know, saying so does not make it so; actions must be consistent with declarations, or the current situation will only get worse. We must address the implications of proposed initiatives and applications and learn from the results of our actions. Observers and critics will quickly determine if our pronouncements are real and substantive or mere window dressing for business as usual.

Questions come before answers, problems before solutions, and why before how. More thought must go into clarifying and agreeing upon the problem before we design solutions. (Recall Severeid's rule: the cause of most problems is solutions!) There is frequently a tendency to focus on a technical fix—the how to—rather than the why. People will not be able to deal with details of how to solve a problem until they understand that there is a problem that needs a solution. Selling the problem is necessary before people will accept a solution, and what people are most interested in is the end product, not the tools used to achieve it. Tools are means to ends. We need to understand and agree on the ends desired in order to select the appropriate means for achieving them.

The process must be open and fair. Following from the above, not only must we avoid confusing the means with the ends and inputs with outputs, but we must focus on the process as well as the end point. For example, the process of planning is often more important than the plan itself; and the process we use to make decisions can be the key to whether the decision itself is understood and accepted. Sometimes what we learn along the way may lead us to a previously unknown destination. For any alternative model of forest management to succeed—New Perspectives, ecosystem management, or whatever tomorrow holds—it will be necessary that it feature an open process that fairly considers all points of view and fosters mutual learning and an adaptive approach.

People will not support what they do not understand and cannot understand what they are not involved in. Many professionals bemoan the seeming lack of understanding the public has for natural resource issues. In many respects, this is probably true. But professionals do not understand the public very well either. And the seemingly whimsical nature of public opinion about forestry issues compounds this concern. This may be a result of an increasingly urbanized and diverse population whose basic information comes from the media. It may also be due to ineffective approaches employed by professionals to help people understand one another and the complex world of natural resources. To change the situation, new models of management must require that public education and involvement processes become truly participatory, with the public an active partner in mutual learning, understanding, and action.

People who do not share your values are not necessarily your "enemies." One of the most disturbing characteristics of the debate over natural resources in the United States is the shrillness of the dialogue and the vilification of people of opposing values. Loggers, foresters, urbanites, scientists, bureaucrats, and environmentalists have all been painted as villains, depending on the point of view of the person speaking or reporting. Such a tactic makes hollow the claim by the same people that a middle ground or common ground is needed. Polarization may be useful in establishing points of view in ten-second sound bites or quickly making a point on the editorial page, but it is disruptive to achieving the consensus so many interests claim they want.

Change is the only constant. Accept it. People seeking stability in the relationship between natural resources and societal values, uses, and demands are likely to be disappointed if the past (and present) is a prologue to the future. The rate of change may increase; the nature of the pressures faced may vary. As noted earlier, we must recognize the development of New Perspectives and ecosystem management as a stage in an evolving philosophy and approach to forest management. But unless we learn along the way, we may find that what is a new perspective today may be part of the problem tomorrow. We must continuously and carefully monitor the situation and adapt as necessary and appropriate. Hopefully, an evolutionary process may preclude a revolution.

Leadership and vision are needed. Before we can hope to achieve our goals, we need a shared vision of what they should be and how we will attempt to achieve them. That is the basic task of leadership. And leadership is not the exclusive domain of top managers of hierarchical organizations. We need to cultivate and reward leadership from below as well as from above, from within the ranks of our organizations, institutions, and communities, and from outside traditional structures. Professionals must seize the opportunity to improve the responsiveness of forest management to public needs and desires, to manage for a wider variety of values and uses, and to integrate physical, ecological, and social considerations into planning, management, education, and research. If professionals cannot find creative solutions from within, ''solutions'' most certainly will be imposed from without.

Dream about what could be; it may well come true. One of the tragedies of modern society is how quickly most children lose the ability to dream about what could be and how to make their dreams come true. Conforming and avoiding risks become the norm. Boundaries, roles, rules, and politically correct thinking are the stuff of nightmares, not dreams. All of these constrain discourse and creativity in the search for solutions to vexing problems. Implementation of New Perspectives and an ecosystem approach to forest management require us to stretch beyond these limits to find creative approaches to resolve the increasingly complex and contentious problems we face. In dreamtime, rules are meant to be broken. If we will risk breaking through these seeming limitations, the most interesting things may become possible.

A dreamer is one who can only find his way by moonlight
and his punishment is that he sees the dawn
before the rest of the world

Oscar Wilde

ENDNOTES

1. Paper presented at the Second Canada/U.S. Workshop on Visitor Management in Parks, Forests, and Protected Areas, Madison, Wisconsin, May 13–16, 1992.
2. The Forest Ecosystem Management Assessment Team (FEMAT, 1993) added two categories to this classification system—health values (medicines) and security values (sense of social continuity and heritage).

REFERENCES

Clark, Roger N., and Perry J. Brown. 1990. The emerging web of integrated resource management. In *Proceedings of the 1990 IUFRO XIX World Congress,* Montreal, Canada 6:24–33.

Clark, Roger N., and George H. Stankey. 1991. New forestry or new perspectives: The importance of asking the right question. *Forest Perspectives* 1(1): 9–13.

DePree, Max. 1989. *Leadership is an art,* 148 pp. New York: Dell Publishing.

Dunlap, Riley E., J. D. Grieneeks, and Milton Rokeach. 1983. Human values and pro-environmental behavior. In *Energy and material resources: Attitudes, values and public policy,* edited by W. David Conn, 145–168. Boulder, Colorado: Westview Press.

FEMAT. 1993. Forest ecosystem management: An ecological, economic, and social assessment. Report of the Forest Ecosystem Management Assessment Team. Portland, OR: U.S. Department of Agriculture, U.S. Department of Interior, and others.

Koch, Niels Elers. 1990. Sustainable forestry: Some comparisons of Europe and the United States. In *Sustainable forestry: Perspectives for the Pacific Northwest.* The Starker Lectures, edited by James R. Boyle, 41–53. Corvallis: College of Forestry, Oregon State University.

Kuhn, T. S. 1970. *The structure of scientific revolutions,* 2d ed. Chicago, Illinois: University of Chicago Press.

Stankey, George H., Perry J. Brown, and Roger N. Clark. 1992. Allocating and managing for diverse values of forests: The market place and beyond. Paper presented at the IUFRO International Conference ''Integrated Sustainable Multiple-Use Forest Management under the Market System,'' Pushkino, Moscow Region, Russia, September 7–12, 1992.

Stankey, George H., and Roger N. Clark. 1992. *Social aspects of new perspectives in forestry: A problem analysis,* 33 pp. Gifford Pinchot Institute for Conservation Monograph Series. Milford Pennsylvania: Grey Towers Press.

Wondolleck, Julia M. 1988. *Public lands conflict and resolution: Managing national forest disputes,* 263 pp. New York: Plenum Press.

Chapter 6

Ecosystem Management: A New Perspective for National Forests and Grasslands[1]

Hal Salwasser

Major changes are under way in how America's federal lands and resources are managed. If you read the papers or watch TV reports, you have seen symptoms of the change: conflict, protest, anger, name calling, lawsuits, agency bashing, science bashing, politician bashing, and so on. Driving it all is a problem that becomes more acute with each passing year: more people competing for increasingly scarce resources. The needs and wants of these people create a great challenge for resource managers: we cannot satisfy all the wants. But someone, somewhere has to work on satisfying the needs. Therefore, beneath all the rancor and conflict, substantive change is occurring.

ECOSYSTEM MANAGEMENT EMERGING

In public land and resource management, a new perspective is emerging. It is more focused on ecological principles and more attuned to public values and expectations. The USDA Forest Service calls this new view "ecosystem management." The name is not as important as the principles and how they are used.

To a degree, an ecosystem perspective is new—at least some of the science and technology are new. To a degree, it is also old: the ideas of holistic thinking, land stewardship, and keeping people and the land in harmony have been around for a long time.

Some Definitions

An *ecosystem* is the complex of a community of organisms and the environment working together as an integrated unit; that is the concept. Ecosystems are also real things: they are places like forests and wetlands where plants, animals, soils, waters, climate, people, and the processes of life work together to support life. Ecosystems respond to inputs of energy and how people treat the land. They are always changing, whether people cause the change or not.

Ecosystems do not have absolute or permanent boundaries, though we draw lines around different kinds of places for our convenience. They are all interconnected and are influenced by people directly or indirectly. Things move around in ecosystems over space and time. Every ecosystem is a subset of a larger system. Forests are ecosystems, as are ponds, lakes, rotting logs, rangelands, and estuaries. The Northern Rockies is an ecosystem. So is North America. So is the planet.

Landscapes are large ecosystems that are composed of many smaller ecosystems. They have distinctive patterns of habitats, physical features, and human communities and cultures that make one kind of landscape different from another. Landscapes are bigger than watersheds and smaller than regions. They are what you see when you stand on a scenic vista or fly at low levels in an airplane: twenty thousand to one hundred thousand acres is the approximate size of a landscape. Landscapes often involve multiple ownerships.

Biodiversity is the variety of life in ecosystems. It includes the variety of genes, species, plant and animal communities, and the many processes through which they are interconnected through space and time. The diversity of species and ecological processes is what keeps ecosystems healthy and productive. Therefore the desired result of ecosystem management is healthy, productive ecosystems and healthy, productive human communities that depend on them.

Ecosystem management is the synthesis and skillful use of knowledge and actions in the stewardship of ecosystems to encourage desired conditions of environments, economies, and human well-being. Communities of people must define the desired conditions and set the course for their attainment.

The ecosystem concept views land as a whole and stresses the maintenance of all its parts and processes in good working order, including the human parts. The ecosystem management policy is founded on the original purposes for reserving public lands in this nation: to protect the land and make sure it provides a wealth of benefits for future generations. However, ecosystem management will not make much difference if humans do not temper their accelerating influence in the biosphere.

What Ecosystem Management Is Not

There has been much discussion and debate about ecosystem management during the 1990s. So, let me tell you what it is not. First, it is not just another name for the way things have always been done. For example, it is not just another name for wilderness preservation, on the one hand, or unsustainable resource extraction on the other. These old choices are not globally responsible when taken to the extremes that some propose. Thus ecosystem management is an alternative that includes options to preserve some places and exploit others with an eye toward a balance of environmental protection with meeting people's needs for resources.

In the past, we tended to treat each administrative unit or land owner-ship—each park, forest, refuge, or tree farm—as if it were an ecosystem unto itself, an isolated island. Of course, they are not. They never have been. They never will be. Ecosystem management thus tries to bring the true complexity, dynamics, and interconnectedness of lands, resources, and human communities into better focus.

Looking to the Future while Building on the Past

While ecosystem management is an emerging concept, it does not reject the accomplishments of the past. Land management strategies should always seek to blend the "best of the old," e.g., soil conservation, with the "best of the new," e.g., conservation of biodiversity, and stay open to the "yet to come." These strategies should aim to sustain both healthy land and the uses and values of natural resources that support human well-being. The old approaches were good for their time. The richness of forests, wildlife, parks, and natural areas that we have today is a direct result of what people have done for the past century. But times change, and we learn new things every day. One of the lessons of the 1980s was that we cannot command or control nature; volcanoes, wildfires, hurricanes, and floods all reminded us of this. Another was that simpli-fying natural systems is not the best way to deal with an uncertain future. And, as recent fires, insect outbreaks, effects of unchecked ungulate populations, and droughts have shown us, leaving ecosystems to nature's vagaries does not always yield socially desired results.

A single focus for land and resource management, whether for wilderness or resource production, is still useful and needed in some places. But simple models of land and resource management are not by themselves sufficient to meet the challenges we face in the coming decades. If there are going to be more people (25 to 30 percent more by the early part of the twenty-first century), they will have to find a suitable quality of life with less impact on their supporting eco-systems.

Costs of Not Taking an Ecosystem Perspective

Ecosystem management can be a framework for blending many different ap-proaches to land and resource management. It has a place for both wilderness preservation and resource production. But it also allows for many options that are intermediate between these extreme choices. An ecosystem perspective will add a focus on diversity to traditional concerns for preservation and production. But it will also have to address human population growth, the increasing per capita effects of resource use and pollution, new scientific understandings, and greater sensitivity to the conditions of our communities and environments.

The process of making the change to ecosystem management will not be simple. Already it has led to controversy, conflicting views about what sustainable resource management means, and no little amount of political bloodletting. But the costs of *not* making the change are steep. They include continued loss of soil, water, and biotic resources; expansion of globally irresponsible local behaviors; deterioration of parks, refuges, and wilderness areas; disruption of human communities; and disintegration of the common ground that people worked so hard to form over the past century in North American conservation.

NEW PERSPECTIVES FOR MANAGING THE NATIONAL FOREST SYSTEM

The good news is that change is underway. An ecosystem perspective is evolving in many places (though it may be called by many different names, such as stewardship, holistic resource management, and sustainable forestry). There are several aspects of how the ecosystem perspective is evolving in the Forest Service.

The shift toward ecosystem management in the national forests and grasslands started with forest restoration in the early part of this century and with attention to snags, riparian areas, and endangered species several decades ago. Ecosystem concepts influenced Forest Plans in the 1980s. The 1990 Resources Planning Act (RPA) Program (USDA Forest Service, 1990), New Perspectives projects, and various resource initiatives furthered the shift. Now ecosystem management has been adopted as the future direction for stewardship of the national forests and national grasslands (Robertson, 1992; Thomas, 1994).

The Basic Policy Question

Behind all the furor over the future of national forests and grasslands, Americans are grappling with a fundamental policy question for public lands and natural resources. Given the economic, environmental, and social needs and aspirations of our people and given the unique or differential capabilities of all lands to meet those needs and aspirations, what should be the special roles of national forests and grasslands in our communities, economies, and environments?

This question, a variation on Clawson's (1975) "Forests for Whom and for What?," can be framed for any land ownership. It has important corollaries:

- How and by whom should desired roles for national forests be carried out?
- Are new economic or environmental policies, standards, or incentives needed?

- How should those who most directly benefit from these roles pay for the benefits or compensate those whose interests may be disfavored?

Making Tough Choices

The dilemma we face as a society is how to choose from among all the possible benefits that lands such as national forests and grasslands can provide and how to pay for our choices. The current roles of these lands in our communities, economies, and environments are enormous. Over time, they will probably become even greater. For example, national forests and grasslands hold major U.S. market shares of many vital resources, including

— Wilderness areas
— Native vegetation diversity
— Vertebrate species diversity
— Wild and scenic rivers
— Softwood saw timber and furniture-grade hardwoods
— Fisheries, including half the trout and salmon habitat in the nation
— Watersheds
— Outdoor recreation
— Minerals and energy
— Habitat for big game and endangered species

The problem faced by stewards of the national forest system is that our growing population wants more of all these things from their forests and grasslands. Unfortunately, more of everything is not possible. There are limits to the ability of any land area to provide different benefits. There are trade-offs between conflicting uses and values.

National forests and grasslands could produce more wood (to help meet growing domestic and foreign needs). They could provide more recreation (to enhance local economies and people's leisure time). They could focus on the recovery of endangered plant and animal species and the conservation of biological diversity. They could provide more forage for range-fed livestock. They could provide more access to minerals and energy (to reduce foreign dependencies, which are staggering for many materials). They could become the nation's foremost outdoor laboratory and classroom for urban people to learn about what it means to live in harmony with land and resources.

But national forests and grasslands cannot do more of all these things at the same time or in the same place. Choices have to be made. Which uses and values should be favored? Who will pay for benefits? Who will compensate those displaced by these choices? How can we blend management for different values and uses to get the best mix while sustaining productive and diverse ecosystems?

New Directions

In the context of these choices, new directions are being taken in how national forests and grasslands are managed. Strategic guidance from the 1990 RPA Program features four themes:

- Increased attention to recreation, wildlife, and fisheries resources
- Increased environmental sensitivity in resource production projects
- Increased research on natural resources and how ecological systems function
- Increased attention to related global issues through research, resource management, technical assistance, and international programs

Local management to meet these new directions is guided by integrated forest plans. Teams of managers, scientists, academicians, and citizens develop and carry out projects under these plans. Some, but not all, of the projects test the application of new knowledge, new technologies, and new alliances with universities and the public. During 1990–1992, more than two hundred such projects were carried out under the Forest Service's New Perspectives Program, about a dozen of which were major research and development programs. These projects led directly to the agency's policy direction on ecosystem management by providing the practical validation for principles and guidelines (Robertson, 1992).

Aims and Purposes of New Perspectives

New Perspectives projects had five primary aims: (1) to learn how to better sustain diverse and productive ecological systems, (2) to integrate all aspects of land and resources management in specific watersheds and landscapes, (3) to improve the effectiveness of public participation in resource decision-making, (4) to build partnerships between forest users and forest managers, and (5) to strengthen teamwork between researchers and managers. Here are some of the lessons learned from the projects.

Ecosystem management means treating land and resources as a whole rather than as a collection of separable parts or processes. It also means recognizing that people are integral parts of the whole. The goals for ecosystem management in any particular place are not intrinsic to the fact that ecosystems exist in that place. They come from the people as expressed through long-standing legal mandates, agency policies, land management plans, local needs, local cultures, and annual budgets.

The aim of ecosystem management in national forests should be to sustain healthy land first, then to provide people with the variety of benefits and options they need and want, consistent with basic land stewardship. The foundation for sustaining healthy, diverse, and productive land is to protect, restore, and main- tain productive soils, clear air, clean water, and a rich biota. However, this is not a new perspective. Leopold (1949) espoused it at mid-century, as did many others before him. But some of the social aspects of ecosystem management are new. Ninety percent of Americans now live in urban or suburban settings. Many of these people no longer know land as a renewable resource. Many professionals involved in natural resource management no longer know these urban people and their needs and values. To say we are out of touch with the land and with one another is an understatement.

To rebuild understanding for land health and resource stewardship and to restore individual and community responsibility for land and resources, we must make participatory decision making and interpretation integral parts of ecosystem management, not luxuries that are available only if time and budgets permit. Experience with New Perspectives projects showed that direct participation in resource management projects was a key to giving people a chance to reconnect with the land, to rebuild their communities, and to educate resource scientists and managers.

Wholes in Addition to Parts

Managing ecosystems to sustain diversity as well as resource productivity is different from how wildlands and natural resources have been managed for the past four decades. Since the 1950s, resource management disciplines evolved from an agricultural philosophy of sustained-yield resource production. Foresters, wildlife habitat managers, fisheries managers, farmers, and rangeland managers all managed ecosystems to produce and sustain the yields of desired products, often commercially valuable resources such as timber, but sometimes amenity resources as well, such as elk or rainbow trout.

Focusing only on outputs of products usually leads to simplified ecosystems. One example is where clearcutting is followed by plantations of desired trees and treatments to enhance wood growth. To meet people's needs for resources, such product-oriented management is necessary in some places. It is certainly more viable economically and ecologically than the old exploitation models of resource use: cut out and get out. But it cannot and should not be applied everywhere. It is most suited to low-elevation sites on flat terrain with deep soils, high rainfall, long growing seasons, and proximity to markets. Not all wildlands are like this, certainly not most of the lands in the National Forest System. Much of the land in the National Forest System is steep and at high elevation, with shallow soils and short growing seasons. Such lands cannot produce high yields of commercial products in an environmentally acceptable or economically feasible manner. The main value of these lands is to protect watersheds and sustain the natural processes of planetary health. Resource uses and products are thus not the only goals for ecosystem management. But where needed resources can be produced without impairing the health and productivity of the land, and where it makes good economic sense to feature certain uses and products that people want, then resource production is a valid purpose of ecosystem management.

Blending Compatible Uses and Goals

One aim of ecosystem management is to blend management goals and their practices at multiple geographic scales so that overall land and resource management better sustains the health of the land and the diversity of purposes for which it is being managed. This requires high levels of cooperation and coordination between resource disciplines and between different administrative units or land ownerships—interdisciplinary and interdependency are vital features of ecosystem management.

A Foundation

Land and resource managers are still in the early stages of learning to use an ecosystem perspective. Most of the technical details—ecological, economic, and social—are still being developed by field managers, researchers, and citizens. But the foundation and framework are taking shape. For example, we learned from ecological research and New Perspectives projects that sustainable management of ecosystems will have three essential attributes:

1. It will sustain ecological diversity and renewability at multiple geographic scales and over long periods of time (what some might call ecosystem integrity or land health).
2. It will be economically feasible, ideally measured in real and full costs and benefits.
3. It will be socially responsible and responsive to people's needs.

If any one of these factors is too far out of balance with the others, desired ecosystem conditions will not be sustainable. It would be like taking the fuel, heat, or oxygen away from a fire to make the flame go out. You might think of this "triangle" of environment, economy, and community as the foundation for ecosystem management.

Framework Principles

We found that three principles were useful in guiding the development of ecosystem management goals, which are essentially the beginning of a framework for ecosystem management.

First, protect land health by ensuring that productive soils, clear air, clean waters, and a diverse biota sustain what Aldo Leopold (1949) called the "integrity of the land community."

Second, meet the basic needs of people who depend on the land for subsistence, livelihood, and spiritual development in order to sustain a rich diversity of human lifestyles and cultures—within the capacity of the land to maintain its integrity and meet those needs.

Third, develop resource uses and products that contribute to the well-being of human communities and economies—within the capacity of the land to maintain its integrity and meet the basic needs of dependent peoples.

Obviously, balance and equity are crucial to how these principles are used. Too much emphasis on resource uses and products compromises our ability to protect the land. On the other hand, too much emphasis on protecting the land can lead to undesired ecological conditions (such as those that occur with the exclusion of natural fires) or reduces our ability to meet people's needs for resources. Even worse, too much protection of resources in domestic ecosystems can shift the effects of resource production to other nations, where environmental protection is weak. There are tough choices to be made and a balance to be struck and continually refined. Such is reality. Such is the difficulty of keeping people and land in harmony.

If these principles are followed and kept in reasonable balance, the result should be fairly close to what Gifford Pinchot called the "greatest good for the greatest number in the long run" and what the World Commission on Environment and Development (1987) called "sustainable development."

The Role of Research

Research will play a new and significant role in ecosystem management, including the use of scientific methods in designing management so that monitoring can facilitate periodic adaptation to new knowledge. But there will be no unique or scientifically perfect answers to how the balance of goals and practices for ecosystem management should be struck. People's values, preferences, and aspirations all enter the policy-making arena. The role of science in ecosystem management is to define the boundaries of what is possible and to shed light on how to best attain a desired set of conditions, uses, or services. The role of science is to help people understand the estimated costs, benefits, and consequences of alternative courses. It is not the role of science to make people's value judgements for them. This would degrade the objectivity and utility of the sciences. It would turn science into just another secular religion. Further, in ecosystem management the social, biological, and physical sciences must be integrated to reflect the complex and integrated reality of how ecosystems function.

Management Guidelines

In developing ecosystem management as an operating policy, some working guidelines evolved in the New Perspectives projects.

Diversity and Sustainability Diversity is key to adaptability, and adaptability is key to sustainability. Therefore, ecosystem management should work within nature's range of variation, capabilities, and processes to maintain as much diversity as possible and to minimize the energy costs of management. Sustain ecosystems with as much of their natural parts and processes in good working order as possible. Be especially careful with soils and waters and in sensitive areas such as wetlands, riparian zones, fragile sites, and rare species' habitats.

Dynamics, Complexity, and Options   Ecosystems change, everywhere and all the time, including what the people in particular ecosystems need and want from the land. Therefore, think about scale effects, both spatial and temporal, at least one scale higher and one scale lower than what you are working on and at least for one generation into the future (lest you think this is asking for too much, some Native Americans try to think ahead for seven generations). A simple motto might be "think complexity, model simply, and maintain options."

Desired Future Conditions   Focus on the desired end results of ecosystem management and the land-use classes and management practices that will best attain them: the conditions of the land at multiple geographic scales that will best sustain ecological integrity and the desired conditions and flows of resource uses and values needed by human communities.

Coordination   For issues and concerns that cross jurisdictional lines, such as air quality, water flows, migratory fish and wildlife, and commodity supplies, work with others to improve the likelihood of success.

Integrated Data and Tools   Integrate ecological classifications, ecological inventories, geographic information systems, and multiresource simulation models, and use them routinely in landscape-level analyses and conservation strategies.

Integrated Management and Research   Integrate monitoring and research with management to continually improve the scientific basis of ecosystem management.

The Bottom Line   The bottom line in ecosystem management is to have a passion for diversity in ecological systems, in economic systems, and in human communities.

SUMMARY

Ecosystem management is a different way of thinking about how to carry out the management of lands and resources. It blends ecological, economic, and social considerations with a better focus on sustaining whole ecosystems in addition to the flows of resource uses and products people need from the land.

In using New Perspectives projects to develop ecosystem management during 1990–1992, the Forest Service emphasized four themes:

1. Sustaining ecological diversity for healthier land and a wider variety of uses, values, and services.
2. Opening the decision-making process to more effective participation in deciding what to do about public resources.
3. Bringing researchers and resource managers into stronger teamwork relationships on adaptive land and resource management.
4. Improving the integration of all aspects of resource conservation.

By no means did New Perspectives get all four of these themes working together on every Forest Service plan or project. But results of the projects told the agency that the themes were heading in the right direction. Thus, ecosystem management emerged as a policy result of the New Perspectives projects (Robertson, 1992).

The learning process is far from over. In fact, ecosystem management has just begun. We cannot tell yet what all of the results will be. However, it is clear that managing lands for diverse, renewable, and productive ecosystems to serve both this and future generations of Americans will require many changes in how resource management agencies do their business.

Ecosystem management should yield a better future for the health and productivity of wildlands, natural resources, and the human communities that depend on them. It might help us find an alternative to the costly and divisive controversies that are currently forcing people to take unrealistic, either/or positions and are damaging our public institutions for resource management. But better public land stewardship alone will not solve all the challenges of maintaining a healthy planet. Regardless of how well ecosystem management works, other social and economic actions will be needed to bring people and land into a better harmony.

CONCLUSION

The Forest Service mission is to care for the land and provide people with needed services and resources. What this means has changed over the past century. Management practices eventually reflect emerging views, though the time lag is often longer than some would like. New views embodied in ecosystem management do not reject the accomplishments or utility of old concepts and practices that still serve the agency's mission. In fact, the new views build directly on the foundation established by previous accomplishments.

Because of what prior generations of conservationists, educators, scientists, and resource managers created, our generation has the ability to search for better ways to sustain diverse and productive ecosystems for the uses, values, and services that people want and need now and in the future. We are most fortunate to have this chance. People in other nations who did not make the sacrifices and investments in conservation that Americans did for the past hundred years do not have such a heritage.

Conservationists, politicians, and resource managers now have an obligation to not let our current environmental, economic, and social ills destroy this heritage. If we do, the integrity of our land and human communities will suffer. Options for the well-being of future generations will decline. Our spirit of community responsibility will be ripped asunder.

An ecosystem perspective can help us chart a different course if we develop it together, openly, and with a commitment to heal the wounds, renew the land, rebuild our communities, and restore a common ground for globally responsible conservation.

ENDNOTE

1. This paper was delivered while the author was Director of New Perspectives for the USDA Forest Service, Washington, D.C. Dr. Salwasser was the Boone and Crockett Professor of Wildlife Conservation at the University of Montana from September 1992 to June 1995. Subsequently, he was Regional Forester for the Northern Region USDA Forest Service from June 1995 to October 1997, and is currently serving as the Pacific Southwest Research Station Director.

REFERENCES

Clawson, M. 1975. *Forests for whom and for what?* 175 pp. Baltimore, Maryland: Johns Hopkins University Press.

Leopold, A. 1949. *A Sand County almanac and sketches here and there.* 228 pp. New York: Oxford University Press.

Robertson, F. D. 1992. Ecosystem management of the national forests and grasslands. Memo to Regional Foresters and Station Directors, USDA Forest Service, Washington, D.C., June 4, 1992.

Thomas, J. W. 1994. *The Forest Service ethics and course to the future.* FS 567, 9 pp. Washington, D.C.: USDA Forest Service.

USDA Forest Service. 1990. *The Forest Service program for forest and rangeland resources: A long-term strategic plan.* Washington, D.C.: USDA Forest Service.

World Commission on Environment and Development. 1987. *Our common future.* 400 pp. New York: Oxford University Press.

Chapter 7

Toward a Better Understanding of Human/Environment Relationships in Canadian National Parks[1]

Elaine Nepstad

Per Nilsen

The purpose of this chapter is to present a general framework within the concept of ecosystem management for viewing human/environment relationships in Parks Canada. It is developed from a literature review of concepts of ecosystem management and of existing approaches to understanding human/environment relationships. There are many ways of viewing human/environment relationships; the framework presented is one that applies to Parks Canada.

SUMMARY OF ISSUES FROM LITERATURE REVIEW

The literature review points to several issues that currently face those embracing ecosystem management.

We are in the midst of paradigm shifts that concern our views of science and nature. Our understanding of what science is about is changing from a rational, strictly objective practice to one that is viewed as an expression of the culture in which it occurs. The view of nature is changing from mechanistic to holistic, implying that one cannot tinker with one part without affecting the others. For resource managers, these changes indicate a shift toward ecosystem management and collaborative decision making (Cortner and Moote, 1992). However, institutions that are still largely structured to administer policies based on the traditional rational approach will require administrative shifts along with conceptual shifts for ecosystem management to occur (Mitchell, 1991).

Humans must be considered an integral part of an ecosystem, especially at a time of increasing urbanization and detachment from the natural environment. Traditionally, people have been viewed as external to the environment, exerting a disruptive influence on it. An emerging framework places people in the ecosystem, where they interrelate with other components (Stankey and Clark, 1992).

As a result of these paradigm shifts and the adoption of an ecosystem approach, there is a need for integrated studies of people and the environment. To date, people and their relationships between settings, activities, experiences, and benefits are given little attention in ecosystem research, planning, or management, even though they are recognized as being important. Although long pursued in both physical and social science, academic work in this field still remains fragmented and disciplinary, a mode duplicated in the work world.

The implementation of an ecosystem management approach that includes people requires an understanding of values held by various segments of the population toward landscape/seascape, as well as values held by management agencies (Stankey and Clark, 1992; Field and Burch, 1988). Management decisions made without an understanding of these values, for example, the values of resident peoples toward the landscape/seascape, are increasingly ineffective (West and Brechin, 1991).

PROPOSED FRAMEWORK FOR UNDERSTANDING HUMAN/ENVIRONMENT RELATIONSHIPS IN THE CANADIAN PARK SERVICE

Why a Framework?

The purpose of this proposed framework is to illustrate one view of human/environment relationships associated with national parks, a view that specifically highlights the need to consider the human component in ecosystem management. Traditionally, only the physical and management components have been the focus of research, planning, and management of protected areas, but it is increasingly recognized in scientific literature, in federal government policy, and in practice that *inclusion of the human component is crucial to protection and preservation of heritage resources.* It is hoped that this framework helps to illustrate the complexity of managing national parks, broaden the discussion of ecosystem management, and facilitate dialogue between those involved in the preservation and presentation functions of fulfilling Parks Canada's mandate.

The framework proposed here for illustration and discussion purposes is drawn from the work of Firey (1960), Stankey and Clark (1992), and others,

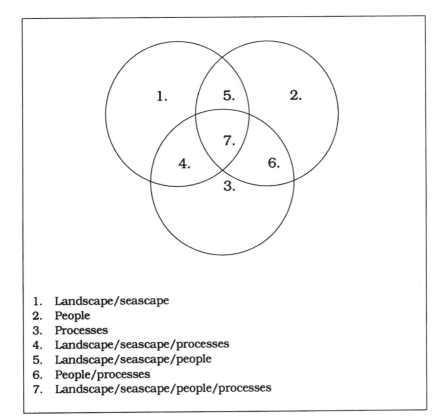

1. Landscape/seascape
2. People
3. Processes
4. Landscape/seascape/processes
5. Landscape/seascape/people
6. People/processes
7. Landscape/seascape/people/processes

Figure 1 Human/environment relationships framework. Adapted from Stankey and Clark (1992).

and is based on the three dimensions commonly found in the literature. It is a simplified picture of complex phenomena; its strength lies in providing a structure for viewing the components and their relationships (Figure 1).

Components of the Framework

The three components in the proposed framework are identified as (1) landscape/seascape, (2) people, and (3) processes. It is recognized that humans are part of both the people and the processes components and that the view of the landscape/seascape is unavoidably a human perspective.

Landscape/seascape represents a human conception of the physical world. It includes the biotic and abiotic elements, for instance, water, air, rocks, flora, and fauna, as well as the symbolic meanings the physical world holds. Seascape is explicitly included in recognition of the need to conserve marine areas and to begin to understand the relationships between seascapes, people, and processes. The people component includes their distribution, values, organization, and behavior. Processes include both ecological processes and those social-cultural processes that influence relationships among people and between people and places (Stankey and Clark, 1992).

Relationships between the Components

One of the properties of ecosystems is that there is an organized connection between parts, but "everything is not connected to everything else" (Agee and Johnson, 1988, 5). Each component can be looked at separately, but perhaps of more interest for ecosystem management purposes is the relationships between components. Four areas of relationships are represented in the framework: (4) processes and the landscape/seascape; (5) landscape/seascape and people; (6) people and processes; and (7) relationships that involve all three components.

Levels of Organization of the Components

Each of the components can be looked at from different levels of organization. For instance, people can be viewed individually, as families, as informal groups, as formal organizations, and as citizens of a country. The following discussion, while not comprehensive, provides examples of levels of organization within Parks Canada.

Landscape/Seascape   A biogeographical and thematic classification of landscape/seascape is used by Parks Canada systems planning (Canada Parks Service (CPS), 1984; Mondor, 1990, 1992). In this system, Canada is divided into thirty-nine terrestrial natural regions and twenty-nine marine natural regions based on biogeographical characteristics. The marine natural regions cover the three coasts of Canada and the Great Lakes. During the systems planning process, potential park areas are identified on the basis of their representativeness of these regions (CPS, 1991a).

This system is one of many frameworks for biogeographical classification (Mondor, 1990). The ecozone system adopted within the Canadian government for reporting purposes (e.g., Canada, 1991) divides Canada into fifteen large and generalized units with the ecozone being the largest (Wiken, 1986). The units become increasingly more homogeneous and detailed as the subunits within each ecozone decrease in size, for example, 15 ecozones, 46 ecoprovinces, 177 ecoregions, and so on down to eco-elements (Wiken, 1986; Mondor, 1990).

The delineation of ecozones and the biophysical system of park systems planning are based on similar criteria; their differences lie in the purpose for which they are used. Ecozones are considered too large to use as a basis for the creation of national parks because of the difficulty of selecting representative samples of such large areas, but to contribute to state of the environment reporting, the parks system has been presented in these terms (Eidsvik et al., 1988).

Political boundaries are another way of organizing landscape/seascapes, portioned into parks, regions, provinces, and countries. These boundaries do not generally coincide with ecosystem boundaries, which range in level from specific sites to watersheds to bioregions and on up to a global level. For instance, a watershed may run through one or more provinces and across international boundaries. These divisions of the landscape/seascape present extraordinary challenges to the management of protected areas; the Man and the Biosphere Program (MAB) and the International Joint Commission on the Great Lakes (Francis, 1991) are two examples of organizational strategies attempting to reconcile the divisions these boundaries create.

Park boundaries present special challenges to ecosystem management. Parks can no longer be managed as isolated from the surrounding region. Many of the national parks are surrounded by agriculture (e.g., Riding Mountain), close to heavily populated areas (St. Lawrence Islands), or threatened by nearby resource extraction (Northern Yukon). The political boundaries of parks seldom match the boundaries of the ecosystem of which they are a part, e.g., the Crown of the Continent area, of which Waterton Lakes National Park is a part. It is the larger ecosystem that needs to be included in order to maintain the ecological integrity of the parks (McNamee, 1992), and negotiations of future park boundaries will need to consider this factor.

People   The basic unit of the people component is the individual, but aggregates range from families, groups of associates based on shared interests, organizations (formally bound groups of associates), communities (shared locale and social system), populations (e.g., citizens of a country), and finally, the global community.

People can be grouped in a multitude of ways, depending on one's objective. For marketing purposes, people can be segmented on demographics and socioeconomics, geographics, psychographics, or behavior/motivation (Heron, 1988). The Visitor Activity Management Process (CPS, 1985) and a component of this process, Service Planning (CPS, 1988), has adopted segmentation of park visitors based on behavior/motivation, i.e., Visitor Activity Groups (VAGS).

Another way of organizing this component is by subdividing people into groups, such as citizens of Canada, park visitors, indigenous populations, local residents, and other stakeholders, i.e., resource extractors and environmental advocacy groups. A publication by West and Brechin (1991) is devoted to the discussion of the role of local residents in the successful management of protected areas. Indigenous groups could be viewed as local residents or stakeholders whose involvement in park planning and management is crucial, especially in northern Canada.

Processes   For the purposes of this paper, processes operate at different levels, which can be termed "spheres of influence." For instance, the sphere of influence of government processes ranges from local to regional to provincial to federal and, increasingly, global. The influence of nongovernmental organizations may also range from local to global, for example, from local conservation groups to Greenpeace, which operates globally.

Parks Canada processes operate on three levels—park, regional, and federal, although the processes overlap in many areas. In this case the levels are nested within each other; that is, the regional level includes the park level and the federal level includes the regional. Generally, processes operating at the federal level influence all the regions and the parks; processes operating at the regional level influence the region and the parks within that region; at the park level, processes relate to the operation of the park and its relationship to the surrounding area.

Processes have been classified as follows (Stankey and Clark, 1992):

Prescriptive Processes These include allocation, planning, and management systems. In Parks Canada these are the Systems Planning Process, Park Management Planning Process (PMP), Visitor Activities Management Process (VAMP), Natural Resources Management Process (NRMP), Environmental Assessment and Review Process (EARP), and Business Planning.

Information Processes These include communication, education, marketing, monitoring and evaluation, and research. Parks Canada examples are interpretation and education programs, the Spaces and Species Program, and target marketing as identified in Regional Marketing Strategies and Park Service Plans.

Conflict Resolution Processes These include policy formation, judicial activity, political activity, and mediation. An example is the public consultation process in establishing new parks and in planning for the management of parks.

Social and Psychological Processes These include socialization, assimilation, selective perception, and adaptation. These processes occur within/between groups and/or within individuals and would include, for example, knowledge, beliefs, motives, expectations, and preferences of park visitors.

Ecological Processes Succession, disturbance, and migration are ecological processes. The forest fires in the Four Mountain Parks and Prince Albert National Park and the migration of animals and birds in Point Pelee, Fundy, Wood Buffalo, and Northern Yukon National Parks are examples of the ecological processes that occur in parks and the surrounding regions.

Example from Wood Buffalo National Park

To illustrate the framework, we can use the management issue of the buffalo herd from Wood Buffalo National Park (Figure 2). According to park staff, the

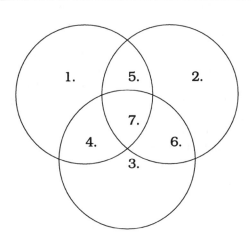

1. **Landscape/seascape:**
    Wood Buffalo National Park
    Peace/Athabasca Delta
    Cattle ranches
2. **People:**
    Native people
    Cattle ranchers
    Scientists
    Government employees
    Environmental groups
3. **Processes:**
    ecological — TB infection of
    bison
    prescriptive — (NRMP, PMP,
    EARP)
    information — communica-
    tion to public
    conflict resolution — what
    to do with the bison
    social and psychological —
    perceptions and values of
    the stakeholders
4. **Relationships between
    processes and land-
    scape/seascape:**
    Provincial land use plan-
    ning — Bennett dam, pulp
    and paper mills
    CPS research, planning and
    management — population
    control of bison and wolves.
5. **Relationships between
    landscape/seascape and
    people:**
    bison — native people
    ranchers
    scientists
    park staff
    park visitors
6. **Relationships between
    people and processes:**
    Native people and
    federal/provincial govern-
    ments — what to do about
    the bison (conflict resolu-
    tion)
7. **Landscape/seascape/
    people and processes:**
    The bison, native people,
    and CPS mangement of the
    herd.

Figure 2 Wood Buffalo National Park example.

population of the herd declined from eleven thousand in 1971 to three thousand
in 1992. This decline in numbers may be attributed to a variety of factors,
including a changing habitat and disease. The herd has been infected with tuber-
culosis (TB), and in 1986 a task force was established to explore ways of eliminating TB, especially since it was considered possible that the bison could pass
the disease on to cattle, humans, and the disease-free Mackenzie Bison herd.
After much deliberation between government departments, a decision was made
to eradicate the herd. The recommendation, supported by the Federal Environmental Assessment Review Office, was vigorously opposed by native groups,
conservation groups, and the park staff. At this point, the issue has not been
resolved and has gone into second review. Due to public outcry, a Northern
Bison Management Board has been established.

1   The *landscape/seascape* component of this issue is the park and the
     surrounding area that affects it, for instance, both the Peace and Atha-
     basca Rivers upstream of the park and the area west of the park used
     for cattle ranching.
2   The *people* involved in this issue include local residents such as native
     people and cattle ranchers; scientists who are studying the bison; a
     range of government people from Parks Canada staff to senior managers
     in the Departments of Agriculture, Health, and Environment; and envi-
     ronmental groups such as Canadian Parks and Wilderness Society,
     Sierra Club, and the Canadian Nature Federation.
3   The *processes* affecting the issue include, for example, ecological pro-
     cesses (TB infection of the bison, secession of the Peace/Athabasca
     Delta), prescriptive processes (NRMP, PMP, and EARP; also resource
     management outside the park such as regulation of waterfowl), infor-
     mation processes (communicating the issue to the public), conflict reso-
     lution processes (how will a decision be reached among those with
     conflicting interests), and social and psychological processes (differing
     values and perceptions held by stakeholders).
4   The *relationship between processes and the landscape/seascape* in-
     clude, for example, the prescriptive processes such as the land/water-
     use planning, which led to the construction of the W.A.C. Bennett
     hydro-electric dam, which affects the water levels of the Peace Delta,
     which in turn, affect the bison; the land/water-use planning that led to
     construction of pulp and paper mills, and the resulting effluent polluting
     the Peace and Athabasca Rivers; and the planning and management
     that led to previous population control of both bison and wolves. These
     are prescriptive processes in that they are the result of the management
     of resources and affect ecological processes and the landscape/sea-
     scape.
5   The *landscape/seascape and people relationships* could focus on peo-
     ple's relationship with the bison, "people" being native people, cattle
     ranchers, scientists, park staff, park visitors, etc.
6   An example of the *people-processes relationship* could focus on one
     group of people, for instance, the native people, and their interaction

with each of the processes. One interaction is their confrontation with the federal government regarding the bison issue (Struzik, 1992), which has to do with conflict resolution processes.

7   The *people-landscape/seascape-processes relationship* is a complex dynamic. An example could be the native people, their interrelationship with the bison, and the effects that Parks Canada management of the bison herd has upon this interrelationship.

Role of Values

Values have to do with what is considered important and given a measure of worth (Stankey and Clark, 1992). They mainly refer to human preferences and are assigned by individuals and society to a wide range of tangibles and intangibles. All of the relationships described in the example are strongly influenced by values held individually and collectively.

People give values to the landscape/seascape, implying that it is of importance and worth to them. Rolston (1988a, 1988b) has identified several values that apply to natural systems. They are life support, economic, recreational, scientific, aesthetic endangered species, historical, cultural symbolization, character-building, diversity-unity, stability and spontaneity, and religious. Stankey and Clark (1992) list values associated with forests that could also be applied to protected areas such as the national parks: commodity, amenity, environmental quality, ecological, public use, and spiritual.

Some values are more easily measured than others. Economic value is regulated by the marketplace and measured by money. Most of the noncommodity values, for example, recreational, aesthetic, and historical, are not so easily evaluated. Other difficulties associated with understanding these values are their distribution across society, i.e., differences between individuals, socioeconomic groups, urban/rural people, and ethnic groups; assessment of changes of values over time; and assessment of worth relative to marketplace values (Stankey and Clark, 1992).

In spite of the difficulties in understanding and assessing values, it is generally recognized that societal values, which include values toward nature, are changing:

> The dominant decision-making values of the industrial age included belief in materialism, private ownership, capitalism, and independence. While still important in Canada today, these perspectives are tempered by increasing emphasis on quality of life, self-fulfillment, humanitarianism, community, and the ecological ethic (Balmer, 1992, 1).

Nash (1989), in his historical review of the change of values toward nature, believes we have entered an era where the rights of man will no longer take precedence over the rights of the natural world.

The trend toward an ecological ethic is reflected in the growing awareness of the environment and the increasing demands to participate in decision making regarding its use. For example, the traditional view that commodity values are the prime importance of land and water resources is being challenged in the Pacific Northwest (Cortner and Moote, 1992), in Temagami, and in the James Bay area of northern Quebec by groups who hold other values for the resource. Another example is a current lawsuit in California, where the Surfrider Foundation is suing Chevron Oil Corp. for building a rock barrier that changed the wave patterns in a bay, spoiling the area for surfing (*Globe and Mail,* March 28, 1992). This group is advocating that the recreational value of the water not be compromised to economic values. The recent focus on ecosystem management and ecological integrity within Parks Canada is a reflection of these societal changes.

The values represented in the Wood Buffalo National Park example can be described at three levels of landscape/seascape: local/regional, national, and international. The possible groups, and values in which they may place greater importance, are listed below:

native people—subsistence, ecological, and spiritual values
ranchers—economic values
conservation groups—environmental values
Parks Canada and other government agencies—protection, public use, and amenity values.

The difficulty of resolving the bison issue in Wood Buffalo reflects the diversity and strength of values and stresses the importance of their consideration in making management decisions.

Issues within the Parks Canada Context

Parks Canada's Guiding Principals and Operational Procedures (1994) (hereafter referred to as Parks Canada Policies) state,

> People and the environment are inseparable. Protection and presentation of natural and cultural heritage take account of the close relationship between people and the environment . . . relationships between people and the environment cannot be separated.

This statement underscores the importance of increasing our understanding of human/environment relations and of integrating this knowledge into the ecosystem management approach to which Parks Canada is committed.

To develop an understanding of the interaction between humans and their environment, Parks Canada needs to increase its scientific knowledge base in this area. Parks Canada has recently reaffirmed its commitment to science. Parks Canada Policies (1994) state,

Management decisions are based on the best available knowledge, supported by a wide range of research, including a commitment to integrated scientific monitoring. Parks Canada requires applied and basic research and monitoring activities to make responsible decisions in its management, planning, and operating practices, as well as to broaden scientific understanding.

A National Task Force on Science and Protection has been established to examine the issues of acquisition, management, and application of scientific knowledge within Parks Canada. Regionally, both of Parks Canada's former Western and Prairie and Northern Regions have produced reports that examine the use of science in management decision making (CPS, 1992a, 1992b). To date, these initiatives have acknowledged the importance of human/environment relations but have not included a comprehensive review of these relationships and the need for integrated scientific research and collaboration among functions.

The knowledge on human/environment relationships acquired through research and practical experiences must be integrated into decision making. Here Parks Canada is not starting from scratch. The framework presented in this chapter and the existing VAMP could be used to integrate the knowledge of people who know the natural resource in day-to-day operations, training, management planning, resource conservation planning, and ecosystem management projects. The application of tools such as the Recreation Opportunity Spectrum (ROS) and the Limits of Acceptable Change (LAC) in test cases to resolve use/ protection issues are also steps toward this end. As recommended by the Task Force on Parks and People (CPS, 1990c, 12–13), ROS and LAC need to be evaluated to determine how they can contribute to VAMP and other Parks Canada planning and management processes. Payne, Carr, and Cline (1997) recently completed an evaluation of ROS through two pilot applications at Pukaska and Yoho National Parks. A comparative analyses of ROS and LAC and other planning tools was completed by Nilsen (1997) and presented at a Limits of Acceptable Change and Related Planning Processes Workshop in May 1997. Additional work is required to improve VAMP through further application and evaluation of the Visitor Activity Concept (CPS, 1991b) and Service Planning (CPS, 1988).

Lastly, the role of the environmental assessment and review process as a tool to assist in improving the management of human/environment relationships needs to be examined and better integrated into existing processes.

Recommendation

Considering the issues posed by ecosystem management and the importance of human/environment relationships in this approach, the authors recommended that Parks Canada establish an interdisciplinary working group. This working group, reporting to the director-general, National Parks, would identify policy, planning, and management issues on human/environment relationships.

One of the first tasks of the working group would be to develop an agenda for the examination of human/environment relationships. Guidelines for setting such an agenda are suggested below.

Goal The goal of the working group is to increase Parks Canada National Parks acquisition, management, integration, and application of scientific knowledge about human/environment relationships, resulting in improved ecosystem management and enhanced environmental citizenship.

Objectives

1   To determine our existing scientific knowledge base and activities related to understanding human/environment relationships
2   To determine priority human/environment relationship issues that need to be addressed
3   To develop an action plan based on the priority issues
4   To recommend a variety of mechanisms, both internal and external to Parks Canada, that will help accomplish the action plan

Rationale Development of an agenda would enable Parks Canada to initiate the long-term and interdisciplinary effort that is required to become knowledgeable clients in the area of human/environment relations. This initiative would contribute directly to (1) achievement of ecological and cultural resource integrity of special places and (2) fostering of public environmental and cultural heritage citizenship (CPS, November 22, 1992d).

Such an agenda would also contribute to the fulfilment of requirements for the effective management of natural systems as outlined by Woodley and Theberge (1992):

1   clear management objectives
2   sufficient knowledge about the ecosystem
3   sufficient power to ensure necessary actions are taken
4   feedback mechanisms, i.e., monitoring and evaluation

Guidelines To reach this goal, Parks Canada should consider the following guidelines.

1   An agenda needs to have a clearly defined sense of direction, to be focused, and to fit into the agency mission and culture (Van Haveren and Hamilton, 1992).
2   The agenda should be established through collective decision making (Ewert, 1990). For Parks Canada, this could include

—stakeholders, i.e., environmental groups and interested citizens

—staff from park, regional, and national levels of Parks Canada, and from each branch, i.e., natural resources, visitor activities, socioeconomics, architecture and engineering, and program management directorate

—the academic community, from both natural and social science disciplines

3　　Actions emerging from the agenda must provide usable knowledge to protected area managers; that is, the knowledge should be what managers need and available when they need it (Machlis, 1992).

4　　The actions should place as much emphasis on the interpretation and dissemination of results as on the conduct of the research itself (Ontario Ministry of Tourism and Recreation, 1992).

5　　The actions and results of the working group must be integrated into decision making (Machlis, 1992). Appropriate organizational conditions and incentives must exist for this to happen.

6　　The actions and results of the working group must be multidisciplinary, including input from natural sciences, social sciences, and arts disciplines.

7　　The actions and results should increase communication and partnership between researchers, practitioners, and policy makers (Ontario Ministry of Tourism and Recreation, 1992).

8　　Initiatives undertaken by the working group should be rigorous, credible, and scientifically defensible (Ontario Ministry of Tourism and Recreation, 1992).

9　　Undertaking these initiatives and building upon them at the field, regional, and national office level is fundamental to Parks Canada moving toward an integrated approach to addressing ecosystem management problems:

We live in a complex and integrated environment. All creatures, including humans, interact with and depend on each other. They all draw on the materials and energy of the physical environment to obtain food and recycle wastes. They all affect each other's behavior.

In the past, responses to environmental problems paid very little attention to these important inter-relationships. Today, the increasing number and complexity of environmental issues demand that we adopt a more integrated approach (Canada, 1990).

ENDNOTE

1. This is an edited version of Occasional Paper No. 5, ''Toward a Better Understanding of Human/ Environment Relationships in Canadian National Parks, Parks Canada, 1993.'' References used in the original literature review, a summary of which appears in this version, are included in the reference list in order to provide additional useful sources for the reader.

We wish to thank the following for their good advice and/or editorial comments: the late Robert Graham, Department of Leisure and Recreation, University of Waterloo; Steve Woodley, Natural Resources, National Parks; Grant Taylor, Parks Canada (retired); Dr. George Francis, Department of Environment and Resource Studies, University of Waterloo; Gary Sealey, Parks Canada (retired); J. R. Gauthier, National Parks; and to all who listened to progress reports at various times during the development of this project and supported it with their input.

Salt Plains—Wood Buffalo National Park, Northwest Territories, Canada. Photo Credit: Parks Canada.

REFERENCES

Agee, J. K., and D. R. Johnson, eds. 1988. *Ecosystem management for parks and wilderness.* Seattle: University of Washington Press.

Allen, T. F. H., and T. B. Starr. 1982. *Hierarchy: Perspectives for ecological complexity.* Chicago, Illinois: University of Chicago Press.

Altman, I., and J. F. Wohlwill, eds. 1983. *Behavior and the natural environment.* New York: Plenum Press.

Balmer, K., ed. 1992. Our values and attitudes continue to shift—maturing into the "post industrial" paradigm. *Leisure Watch Canada* 1(2): 1–4.

Batisse, M. 1992. Biosphere reserves. Paper presented at Fourth IUCN Congress, Caracas, Venezuela, February 1992.

Bennett, J. W. 1976. *The ecological transition.* New York: Pergamon.

Bennett, J. W. 1980. Social and interdisciplinary sciences in U.S. MAB: Conceptual and theoretical aspects. In *Social sciences, interdisciplinary research and the U.S. Man and the Biosphere Program,* edited by E. Zube. Washington, D.C.: U.S. Man and the Biosphere Secretariat.

Berman, M. 1981. *The reenchantment of the world.* New York: Cornell University Press.

Bird, E. 1987. The social construction of nature: Theoretical approaches to the history of environmental problems. *Environmental Review* 11(4): 255–264.

Bonnicksen, T. M. 1991. Managing biosocial systems: A framework to organize society-environment relationships. *Journal of Forestry* 89(10): 10–15.

Boulding, K. E. 1966. *The impact of the social sciences.* New Brunswick, New Jersey: Rutgers University Press.

Bratton, S. P. 1985. National park management and values. *Environmental Ethics* 7(2): 116–133.

Bruhn, J. G. 1974. Human ecology: A unifying science? *Human Ecology* 2(2): 105–125.

Burch, W. R., Jr. 1988. Human ecology and environmental management. In *Ecosystem management for parks and wilderness,* edited by J. K. Agee and D. R. Johnson. Seattle: University of Washington Press.

Butler, R. W. 1982. Geographical research on leisure: Reflections and anticipations on Accrington Stanley and fire hydrants. In *Perspectives on the nature of leisure research,* edited by D. Ng and S. Smith. Waterloo: University of Waterloo Press.

Buttel, F. H. 1986. Toward a rural sociology of global resources: Social structure, ecology, and Latin American agricultural development. In *Natural resources and people: Conceptual issues in interdisciplinary research,* edited by K. A. Dahlberg and J. W. Bennett. Boulder, Colorado: Westview Press.

Canada. 1990. *Canada's Green Plan.* Ottawa: Supply and Services.

Canada. 1991. *Human Activity and the Environment.* Ottawa: Statistics Canada.

Canada Parks Service. 1988. Amendment to the National Parks Act. Ottawa: Environment Canada, Parks Service.

Canadian Environmental Advisory Council. 1991. *A protected areas vision for Canada.* Ottawa: Ministry of Supply and Services.

Canadian Parks Service (CPS) 1984. *National parks and national marine parks systems planning process manual.* Ottawa: National Parks, National Parks Systems Planning.

———. 1985. *Management process for visitor activities.* Ottawa: National Parks, Visitor Activities Branch.

———. 1988. *Getting started: A guide to park service planning.* Ottawa: National Parks, Visitor Activities Branch.

———. 1990a. *Canadian Parks Service proposed policy.* Ottawa: Environment Canada.

———. 1990b. *The Canadian Parks Service strategic plan.* Ottawa: Environment Canada.

———. 1990c. *Final report of the Parks and People Task Force.* Calgary: Environment Canada, National Parks, Western Regional Office.

———. 1991a. *National parks system plan.* Ottawa: Environment Canada.

———. 1991b. *Visitor activity concept.* Ottawa: National Parks, Visitor Activities Branch.

———. 1992a. *Science in decision making,* vols. 1 and 2. Winnipeg: Environment Canada, National Parks, Prairie and Northern Regional Office.

———. 1992b. Towards a sustainable ecosystem: A strategy to enhance ecological integrity. Calgary: National Parks, Western Regional Office.

———. 1992c (March). Wood Buffalo National Park management plan review. *Program Newsletter.*

———. 1992d. CPS results model, November 22. Ottawa: Environment Canada.

Cartwright, T. J. 1991. Planning and chaos theory. *Journal of the American Planning Association* 57(1): 44–56.

Catton, W. R., Jr. 1984. Human ecology and social policy. In *Sociological human ecology: Contemporary issues and applications,* edited by M. Micklin and H. M. Choldin. Boulder, Colorado: Westview Press.

Checkland, P. 1981. *Systems thinking, Systems practice.* Toronto: John Wiley and Sons.

Christensen, N. L. 1988. Succession and natural disturbance: Paradigms, problems, and preservation of natural ecosystems. In *Ecosystem management for parks and wilderness,* edited by J. K. Agee and D. R. Johnson. Seattle: University of Washington Press.

Cortner, H., and M. A. Moote. 1992. Setting the political agenda: Paradigmatic shifts in land and water policy. Paper presented at the 4th North American Symposium on Society and Resource Management, Madison, Wisconsin, May 17–20, 1992.

Dahlberg, K. A., and J. W. Bennett, eds. 1986. *Natural resources and people: Conceptual issues in interdisciplinary research.* Boulder, Colorado: Westview Press.

Day, T. J. 1992. Organizational implications of ecosystem-based management outline. Document prepared for National Managers Conference, Department of the Environment, Ottawa, March 1992.

D'Elia, C. F. 1988. *The cycling of essential elements in coral reefs.* In *Concepts of ecosystem ecology: A comparative view,* edited by L. R. Pomeroy and J. J. Alberts. New York: Springer-Verlag.

Dorney, R. S., and L. C. Dorney, eds. 1989. *The professional practice of environmental management.* New York: Springer-Verlag.

Ehrlich, P. R., A. H. Ehrlich, and J. P. Holdren. 1973. *Human ecology: Problems and solutions.* San Francisco, California: W. H. Freeman.

Eidsvik, H., J. Carruthers, P. de Grandmont, S. Meis, D-M. Grainger, and B. Aris. 1988. *People and parks: Visits to national parks by ecozone.* State of the Environment Series Report-CPS 1. Ottawa: Environment Canada, Canadian Parks Service.

Ewert, A. 1990. Decision-making techniques for establishing research agendas in park and recreation systems. *Park and Recreation Administration* 8(2): 1–13.

Field, D. R., and W. R. Burch Jr. 1988. *Rural sociology and the environment.* Middleton, Wisconsin: Social Ecology Press.

Field, D. R., and D. R. Johnson. 1987. The interactive process of applied research: A partnership between scientists and park and resource managers. In *Social science in natural resource management systems,* edited by M. L. Miller, R. P. Gale, and P. J. Brown. Boulder, Colorado: Westview Press.

Firey, W. 1960. *Man, mind and land: A theory of resource use.* Glencoe, Illinois: Free Press.

———. 1990. Some contributions of sociology to the study of natural resources. In *Community and forestry: Continuities in the sociology of natural resources,* edited by R. G. Lee, D. R. Field, and W. R. Burch Jr. Boulder, Colorado: Westview Press.

Forman, R. T. T., and M. Godron. 1986. *Landscape ecology.* Toronto: John Wiley and Sons.

Francis, G. 1991. Ecosystem management. Presented at the Tri-National Conference on the North American Experience Managing Transboundary Resources: The United States and the Boundary Commissions.

Franklin, U. 1990. Reflections on science and the citizen. In *Planet under stress: The challenge of global change,* edited by C. Mungall and D. J. McLaren. Toronto: Oxford University Press.

Freeman, M. M. R. 1979. Traditional land users as a legitimate source of environmental expertise. In *The Canadian national parks: Today and tomorrow conference II: Ten years later,* vol. 1, edited by J. G. Nelson et al. Waterloo: University of Waterloo, Faculty of Environmental Studies.

Friedmann, J. 1987. *Planning in the public domain.* Princeton, New Jersey: Princeton University Press.

Geertz, C. 1963. *Agricultural involution: The process of ecological change in Indonesia.* Berkeley: University of California Press.

Globe and Mail. 1992. Beach row centres on worth of wave, March 28.

Goldsmith, E. 1988. The way: An ecological world-view. *The Ecologist* 18(4–5): 160–185.

Goldstein, B. 1992. The struggle over ecosystem management at Yellowstone. In *Science and the management of protected areas,* edited by M. Willison et al. Proceedings of an international conference held at Acadia University, Nova Scotia, May 14–19, 1991. New York: Elsevier Science.

Golley, F. B. 1986. Ecosystems and natural resource management. In *Natural resources and people: Conceptual issues in interdisciplinary research,* edited by K. A. Dahlbert and J. W. Bennett. Boulder, Colorado: Westview Press.

———. 1991. The ecosystem concept: A search for order. *Ecological Research* 6:129–138.

Graham, R., R. Payne, and P. Nilsen. 1988. Visitor activity management in Canadian national parks: Marketing within the context of integration. *Tourism Management* (March): 44–62.

Grzybowski, A., and D. S. Slocombe. 1988. Self-organization theories and environmental management: The case of South Moresby. *Environmental Management* 12:463–487.

Hawley, A. H. 1944. Ecology and human ecology. *Social Forces* 398–405.

———. 1984. Prologue—Sociological human ecology: Past, present, and future. In *Sociological human ecology: Contemporary issues and applications,* edited by M. Micklin and H. M. Choldin. Boulder, Colorado: Westview Press.

Heberlein, T. A. 1988. Improving interdisciplinary research: Integrating the social and natural sciences. *Society and Natural Resources* 1(1): 5–16.

Heron, W. 1988. *Market segmentation: Application to VAMP.* Ottawa: CPS Documentation Centre.

Hills, G. Angus, D. V. Love, and D. S. Lacate. 1970. *Developing a better environment: Ecological land-use planning in Ontario.* Toronto: Ontario Economic Council.

Hollick, M. 1992. Self-organising systems and environmental management. Center for Water Research, University of Western Australia.

Holling, C. S., and S. Bocking 1990. Surprise and opportunity: In evolution, in ecosystems, in society. In *Planet under stress: The challenge of global change,* edited by C. Mungall and D. J. McLaren. Toronto: Oxford University Press.

Kay, J. J. 1991. The concept of ecological integrity, alternative theories of ecology, and implications for decision-support indicators. In *Economic, ecological, and decision theories,* edited by P. A. Victor, J. J. Kay, and H. J. Ruitenbeek. Ottawa: Canadian Environmental Advisory Council.

Kaza, S. 1989. Feedback loops that work. Presented at the Society for Ecological Restoration and Management Annual Meeting (January), Oakland, California.

Kuhn, T. S. 1970. *The structure of scientific revolutions.* Chicago, Illinois: University of Chicago Press.

Lieff, B. C. 1989. Case study: Waterton Biosphere Reserve. In *Proceedings of the symposium on biosphere reserves,* edited by W. P. Gregg, S. L. Krugman, and J. D. Wood Jr. Fourth World Wilderness Congress, September 14–17, 1987. Atlanta, Georgia: U.S. Department of the Interior, National Park Service.

Lopoukhine, N. 1992. Engaging the public in public area management. Presented at the Second Canada/U.S. Workshop on Visitor Management in Parks, Forests and Protected Areas at Madison, Wisconsin, May 1992.

Lovelock, J. E. 1979. *Gaia: A new look at life on Earth.* Oxford: Oxford University Press.

Machlis, G. E., D. R. Field, and F. L. Campbell. 1981. The human ecology of parks. *Leisure Sciences* 4(3): 195–212.

Machlis, G. E. 1992. Social science and protected area management: The principles of partnership. Presented at the Fourth ICUN World Congress on National Parks and Protected Areas, Caracas, Venezuela, February 1992.

Machlis, G. E., and D. L. Tichnell. 1985. *The state of the world's parks: An international assessment for resource management, policy and research.* Boulder, Colorado: Westview Press.

McHarg, I. L. 1969. *Design with nature.* Garden City, New York: Natural History Press.

McIntosh, R. P. 1985. *The background of ecology: Concept and theory.* New York: Cambridge University Press.

McNamee, K. 1992. Out of bounds. Nature Canada 21(1): 35–41.

Merchant, C. 1980. *The death of nature: Women, ecology and the scientific revolution.* New York: Harper and Row.

———. 1987. The theoretical structure of ecological revolutions. *Environmental Review* 11(4): 265–274.

Merriam, G., J. Wegner, and S. Pope. 1992. *Parklands: Parks in their ecological/landscape.* Report prepared by the Carleton University Landscape Ecology Laboratory for Canadian Parks Service. Ottawa: Carleton Institute of Biology, Department of Biology, Carleton University.

Micklin, M. 1984. The ecological perspective in the social sciences: A comparative overview. In *Sociological human ecology: Contemporary issues and applications,* edited by M. Micklin and H. M. Choldin. Boulder, Colorado: Westview Press.

Mitchell, B. 1991. ''Beating'' conflict and uncertainty in resource management and development. In *Resource management and development: Addressing conflict and uncertainty,* edited by B. Mitchell. Toronto: Oxford University Press.

Mondor, C. 1990. The concept of representativeness in the design of protected areas systems. In *Canadian Council on Ecological Areas annual meeting,* J. R. Reid, compiler. CEAC Occasional Paper No. 10. Ottawa: Canadian Council on Ecological Areas.

————. 1992. Appendix V: Planning for Canada's system of national marine parks. In *Marine, lake and coastal heritage: Proceedings of a Heritage Resources Center workshop,* edited by R. Graham. Occasional Paper No. 15. Waterloo, Ontario: University of Waterloo, Department of Recreation and Leisure Studies and Heritage Resources Centre.

Moran, E. F., ed. 1984. Limitations and advances in ecosystem research. In *The ecosystem approach in anthropology,* edited by E. F. Moran. Boulder, Colorado: Westview Press.

————. 1990. *The ecosystem approach in anthropology: From concept to practice.* Ann Arbor: University of Michigan Press.

Nash, R. 1989. *The rights of nature: A history of environmental ethics.* Madison: University of Wisconsin Press.

Naveh, Z., and A. S. Lieberman. 1984. *Landscape ecology: Theory and application.* New York: Springer-Verlag.

Nelson, J. G. 1978. International experience with national parks and related reserves. In *International experience with national parks and related reserves,* edited by J. G. Nelson, R. D. Needham, and D. L. Mann. Waterloo: University of Waterloo, Faculty of Environmental Studies, Department of Geography.

————. 1991. Research in human ecology and planning: An interactive, adaptive approach. *Canadian Geographer* 35(2): 114–127.

Nilsen, P. 1987. Visitor activity management in northern national parks: An exploratory study. Masters thesis, University of Waterloo, Recreation and Leisure Studies.

Nilsen, P. 1997. A comparative analysis of protected area planning and management frameworks. In *Proceedings—Limits of acceptable change and related planning processes: Progress and future directions,* edited by Stephen F. McCool and David N. Cole. Gen. Tech. Rep. INT-GTR: 000. Ogden, Utah: Intermountain Research Station, USDA Forest Service.

Ng, D., and S. Smith, eds. 1982. *Perspectives on the nature of leisure research.* Waterloo: University of Waterloo Press.

Odum, E. P. 1986. Perspective of ecosystem theory and application: Introductory review. In *Ecosystem theory and application,* edited by N. Polunin. Toronto: John Wiley and Sons.

————. 1989. *Ecology and our endangered life-support systems.* Sunderland, Massachusetts: Sinauer Associates, Inc.

Ontario Ministry of Tourism and Recreation 1992. *Recreation research and evaluation strategy.* Toronto: Research and Evaluation Section, Recreation Policy Branch, Recreation Division.

Opie, J. 1985. Environmental history: Pitfalls and opportunities. In *Environmental history: Critical issues in comparative perspective,* edited by K. E. Bailes. New York: University Press of America.

Park, R. E. 1936. Human ecology. *American Journal of Sociology* 42(July): 1–15.

Parks Canada. 1994. Guiding Principles and Operating Policies. Ottawa: Parks Canada.

Payne, R. J., A. Carr, and E. Cline. 1997. Applying the recreation opportunity spectrum (ROS) in two Canadian national parks. Occasional Paper 8. Ottawa, Ontario: Natural Resources Branch, National Parks, Parks Canada.

Payne, R., and R. Graham. 1984. Towards an integrative approach to inventory. *Park News* 20(4): 28–32.

Petulla, J. M. 1985. Environmental values: The problem of method in environmental history. In *Environmental history: Critical issues in comparative perspective,* edited by K. E. Bailes. New York: University Press of America.

Pomeroy, L. R., E. C. Hargrove, and J. J. Alberts. 1988. The ecosystem perspective. In *Concepts of ecosystem ecology: A comparative view,* edited by L. R. Pomeroy and J. J. Alberts. New York: Springer-Verlag.

Regier, H. A. 1990. Indicators of ecosystem integrity. Presented at International Symposium on Ecological Indicators, Fort Lauderdale, Florida, October 14–18, 1990.

Richerson, P. J., and J. McEvoy, eds. 1976. *Human ecology: An environmental approach.* North Scituate, Massachusetts: Duxbury Press.

Rolston, Holmes, III. 1988a. Human values and natural systems. *Society and Natural Resources* 1(3): 271–283.

———. 1988b. *Environmental ethics: Duties to and values in the natural world.* Philadelphia, Pennsylvania: Temple University Press.

Rowe, S. 1990. *Home places: Essays on ecology.* Edmonton: NeWest Publishers.

Ruitenbeek, H. J. 1991. *Towards new fundamentals: Indicators of ecologically sustainable Development.* Ottawa: Canadian Environmental Advisory Council.

Schultz, A. M. 1967. The ecosystem as a conceptual tool in the management of natural resources. In *Natural resources: Quality and quantity,* edited by S. V. Ciriacy-Wantrup and J. J. Parsons. Berkeley: University of California Press.

Shepherd, P. 1967. Whatever happened to human ecology? *Bioscience* 17(12): 891–893.

Slocombe, D. S. 1989. History and environmental messes: A nonequilibrium systems view. *Environmental Review* 13(3–4): 1–13.

———. 1990. Complexity, change and uncertainty in environmental management: The Kluane/Wrangell–St. Elias Region. Ph.D. thesis, University of Waterloo, School of Urban and Regional Planning.

Spooner, B. 1984a. *Ecology in development: A rationale for three-dimensional policy.* Tokyo: United Nations University.

———. 1984b. The MAB approach: Problems, clarifications and a proposal. In *Ecology in practice. Part II: The social response,* edited by F. di Castri, F. W. G. Baker, and M. Hadley. Dublin: Tycooly International Publishing Limited. UNESCO, Paris.

Stankey, G. H., and R. N. Clark. 1992. *Social aspects of new perspectives in forestry: A problem analysis.* Gifford Pinchot Institute for Conservation Monograph Series. Milford, Pennsylvania: Greytowers Press.

Steward, J. 1955. *Theory of culture change.* Urbana: University of Illinois Press.

Struzik, E. 1992. The rise and fall of Wood Buffalo National Park. *Borealis* 3(2): 11–25.

Taylor, D. M. 1992. Disagreeing on the basics: Environmental debates reflect competing world views. *Alternatives* 18(3): 26–33.

Van Haveren, B. P., and L. E. Hamilton. 1992. Strategic planning for natural resources research. Presented at 4th North American Conference on Society and Natural Resources, Madison, Wisconsin, May 1992.

West, P. C., and S. R. Brechin, ed. 1991. *Resident peoples and national parks: Social dilemmas and strategies in international conservation.* Tucson: University of Arizona Press.

Whyte, A. 1984. Integration of natural and social sciences in environmental research: A case study of the MAB programme. In *Ecology in practice. Part II: The social response,* edited by F. di Castri, F. W. G. Baker, and M. Hadley. Dublin: Tycooly International Publishing Limited. UNESCO, Paris.

Wiken, E. 1986. *Terrestrial ecozones of Canada: Ecological land classification.* Series No. 19. Ottawa: Environment Canada, Lands Directorate.

Woodley, S. 1991. Developing ecosystem goals for Canadian national parks. Ottawa: Natural Resources Branch, Canadian Parks Service.

———. 1992. *Ecological stressors affecting Canadian national parks.* Ottawa: Natural Resources Branch, Canadian Parks Service.

Woodley, S., and J. Theberge. 1992. Monitoring for ecosystem integrity in Canadian national parks. In *Science and the management of protected areas,* edited by M. Willison et al. Proceedings of an international conference held at Acadia University, Nova Scotia, May 14–19, 1991. New York: Elsevier Science.

Worster, D. 1977. *Nature's economy: The roots of ecology.* San Francisco, California: Sierra Club Books.

———, ed. 1988. *The ends of the earth: Perspectives on modern environmental history.* Cambridge: Cambridge University Press.

———. 1990. The ecology of order and chaos. *Environmental History Review* 14(1–2): 1–18.

Young, G. L. 1974. Human ecology as an interdisciplinary concept: A critical inquiry. In *Advances in ecological research,* edited by A. MacFadyen. New York: Academic Press.

Chapter 8

Leadership in a Community of Interests

William E. Shands[1]

This chapter addresses two topics: consensus building in natural resources planning and new concepts of leadership with special application to natural resources policy and management. It particularly discusses a flexible approach to public involvement referred to here as "open decision making" and the mobilization of a "community of interests"—for the present purposes, a broad-based constituency concerned about a parcel of land and its associated resources.

Whatever the public resources agency, managers are finding that their constituents are clamoring for greater access to the decision-making process. At the federal level, numerous laws require agencies to involve the public and consider the views of citizens affected by the proposed action. For example, the National Forest Management Act of 1976 requires that the public be involved throughout the forest land and resources planning process. In response, the Forest Service has developed intricate processes for eliciting public comment on policy and program proposals.

Providing increased public access to the process is only half of the equation. Policy makers—namely, people in authority in public resources management agencies—must reappraise how they view themselves as professionals and their roles in decision making.

It is my thesis that resources agencies are trying to overlay increased public involvement on operational cultures that have difficulty reconciling diverse viewpoints. This, I believe, is a major reason for the contentiousness that plagues resources management in the United States today. In other words, you cannot layer public participation onto what is essentially a control management system and expect the public to come out of the planning process happy.

The answer, in my opinion, is a flexible and comprehensive approach to public involvement, or "open decision making" in communities of interests. This approach is being applied to some national forests today and shows promise. Whether or not open decision making in communities of interests is the answer, it is apparent that changes are required in how agencies work with the public.

This fact was recognized by Forest Service Chief Jack Ward Thomas in recent testimony before a congressional committee. Said Chief Thomas, "We must institutionalize collaboration. We must create forums where reasonable people, be they environmentalists, industry representatives, recreationists, or government officials, in short, anyone with a stake in the management of our nation's forests, can come together to discuss issues, learn from one another, and work toward a consensus on how forests should be managed." He added, "I believe that a common ground can be found, one that a majority of Americans will support" (Thomas, 1994).

What is the situation? We know that controversy and conflict over resource management decisions are increasing in intensity. Moreover, there is a feeling that public resource management agencies have lost direction. A perceptive academic observer of resource management agencies writes:

> The profession of resources management possesses neither the broad public support nor a clear sense of mission and future vision, both characteristics that had distinguished it for more than half a century (Wondolleck, 1991).

In a conference on leadership in natural resources convened by the Pinchot Institute for Conservation, a high-level Forest Service official reviewed some of the current controversies and his agency's responses and observed that "something isn't working right."

In 1990, as part of an intensive Forest Service critique of national forest planning, I directed a Conservation Foundation–Purdue University team that asked people around the country what they thought of the national forest planning process as it had been carried out by the Forest Service over the past decade. What we heard was disturbing.

For example, in Missoula, Montana, I facilitated a small-group discussion focused on how forest planning could be improved. The group consisted of about ten people representing just about every forest user interest—wilderness advocates, cattlemen, miners, the timber industry, and local officials. While they disagreed on some issues, they agreed on many things—more than one might expect. They spoke politely to each other and listened patiently to others' views—even when it was clear that they did not agree.

Toward the end of the session I commented on the way the discussion had gone and asked them how often they got together to talk like this. (I had assumed that they all knew one another, since they were from the Missoula area.) They looked at me with some amazement at my naivete. In fact, I was told, this was the first time they had ever been together in this kind of setting. Most had never met each other face-to-face. Typically, their dialogue—if you could call it that—was carried out in the press or formal hearings in terms far less civil than had been the case in the discussion group.

As the Conservation Foundation—Purdue team went around the country talking to individuals about their experiences with national forest planning, a model of public participation emerged. It went something like this. The Forest Service would announce that it was initiating the development of a forest plan and call a public hearing to solicit views on issues the plan was to address. This forces interests into hard positions at the very beginning of the process. The planners then retreated to their offices and some months later emerged with a draft plan. Of course, it pleased no one, and the result was further polarization. In due course, a plan was released and greeted with a barrage of appeals. The Forest Service then would call the appellants and say, "Why don't we get together and negotiate?" Perhaps the process was best summed up by one workshop participant, who said that the Forest Service "got the interests all fired up and then ran us head-on into each other." This is a simplification, of course, but close enough to actual experience to raise questions about how the public could be more effectively involved.

In the Conservation Foundation—Purdue report, *National Forest Planning: Searching for a Common Vision* (Shands, 1990), we concluded that the commonly applied model of public participation is too rigid and formal. Instead, we advocated what we termed open decision making, in which various interests work together with the Forest Service to solve problems.

Our report set out criteria for this joint problem-solving approach.

- First, encourage the sharing of information.
- Second, encourage a frank exchange of views among all interests, especially before views harden. This should help define problems.
- Third, help identify opportunities for joint problem solving.
- Finally, make it clear how the decision was reached.

I make no claims to having invented this approach. Variations have been used successfully on several national forests, notably the Green Mountain in Vermont and the Bridger-Teton in Wyoming, and in the development of a wilderness plan for the Bob Marshall Wilderness.

What our report did, however, was spell out the process in some detail and, it is to be hoped, legitimize it. That there is a need for theory in support of this kind of public involvement is attested to by findings of two researchers who studied public involvement in planning on thirteen national forests. There was, they found, a low level of public involvement during the middle stages of forest plan development. One of the reasons, they concluded, was that "there was little agency direction on how to collect public input in a way that was relevant for designing alternatives (Blahna and Yonts-Shepard, 1989).

For the agencies, the message was simple: involve your constituents from the very beginning, strive for mutual education, and encourage others to come up with solutions. This is not alternative dispute resolution as described in the growing literature of environmental negotiation and mediation. I am trained as a planner; this kind of open process is the essence of good planning.

I had begun working with several national forests—the National Forests in North Carolina and the Targhee in Idaho—in the application of open decision making when I was asked to direct a project on leadership in natural resources for the Pinchot Institute for Conservation in cooperation with the Forest Service. The project was inspired by Jeff Sirmon, who recently retired as Deputy Chief for International Forestry. Sirmon believes that the complexity of today's management environment is forcing the forestry profession "to consider new leadership roles to guide itself and society toward a more sustainable future." He continues, " . . . imagine the potential of a decision-making process where a group of people with different backgrounds and points of view come together for a common purpose" (Sirmon, 1991).

Sirmon had become intrigued by an approach to leadership that could be termed the "community of interests" model advanced by Ronald A. Heifetz of Harvard's Kennedy School of Government. The community of interests model casts responsibility for problem solving not on an authority figure, but on the group. Heifetz argues that when confronting difficult policy issues, people must struggle with "their orientation, values, and potential tradeoffs . . . [no] leader can magically do this work. Only the group—the relevant community of interests—can do this work" (Heifetz and Sinder, 1990).

In the community of interests, leadership is not the responsibility of any single individual or institution. Rather, leadership is the sum of the effort of many individuals working within communities, organizations, and institutions. Leadership is shared and exercised by many individuals in many different ways, simultaneously and sequentially.

As I studied Heifetz's work, there was an immediate resonance. I now combine Heifetz's community of interests theory with that of open decision making. I have come to see that agencies must change how they perceive their role and the role of their constituent groups. Typically, agencies strive for greater public involvement while maintaining old institutional cultures of control and one-way flows of information, with decision making tightly held within the agency. This does not work.

Over the last three years, I have worked with three national forests in the application of open decision making in communities of interests. Two of these forests are discussed below—the Targhee in Idaho and the Shasta-Trinity in California—in terms of how the process is working there.

Two years ago, the Targhee embarked on the revision of the forest plan. Key questions being addressed have to do with the regeneration of lodgepole pine, the harvesting of mature Douglas fir, maintenance of habitat for the grizzly bear and elk, livestock grazing and the protection of riparian systems, and off-road use by snowmobiles and four-wheel drive vehicles, among others.

From the very beginning, then Forest Supervisor Jim Caswell invited citizens to participate in the plan revision. Interested citizens are now loosely organized into a Citizens' Involvement Group (CIG). This group has met approximately a dozen times in high school auditoriums, churches, community halls, and in the field. Participants have tramped through the forest looking at timber sales, campgrounds, and elk habitat. They have discussed issues and exchanged views. This summer, they helped write a vision statement for the Targhee—their desired future condition for the forest. And in September, they worked to help the forest staff craft management prescriptions and identify areas where management changes appear to be warranted.

On the Hayfork District of the Shasta-Trinity National Forest, District Ranger Karyn Wood is applying open decision making in communities of interest to the development of an ecosystem management plan for a thirty thousand acre watershed. The Butter Creek watershed had been intensively harvested. Much of it burned in 1987—and the fires of that year are still vivid in the memories of many residents of the Hayfork area. Since spring, the Butter Creek group has met four times. It has developed a future vision of the watershed and has begun to work on big issues, especially the use of clearcutting and herbicides.

Typically, participants in these efforts test each other and the authority figure. Officials have had to learn to just sit down and let participants talk an issue through. This takes time—an issue is rarely resolved in a single meeting, and major issues are revisited continuously. Forest Service officials provide extensive information. And they try to stimulate leadership among the members of this community of interests.

It's hard work. It is relatively easy for the various interests to tell the Forest Service what they want, and then savage the agency when it does not deliver exactly what was demanded. It is much more difficult to sit through meetings, listen to people with ideas different from yours, probe for solutions, and reach agreement.

On the Targhee, the Citizens' Involvement Group is using its collective knowledge to address problems. While there has been some grumbling, they have stuck with it. Dale Pekar, the leader of the Targhee's interdisciplinary planning team, observes that the process is ''like tossing jagged boulders into a rock polisher. People have rubbed elbows in informal settings . . . exchanging views and sharing experiences. People have come to recognize one another as human beings rather than as abstract position statements.''

The community of interests model empowers constituents and this will require adjustments by individuals and organizations. In the community of interests model, Heifetz sees the leader's role not as problem solver, but as one who mobilizes others to solve problems. The challenge now is to stimulate the group to do work: to address problems and come up with solutions. Robert Reich former Secretary of Labor, describes the role of the public manager: ". . . The public manager's job is not only , or simply, to make policy choices and implement them. It is also to participate in a system of democratic governance in which public values are continuously rearticulated and recreated'' (Reich, 1990).

It is clear that the role of the manager changes significantly in these open processes. Rather than acting as an authority, the manager becomes the facilitator of the process with multiple roles: as *catalyst* stimulating the community to work; as *communicator* ensuring that information, views, and facts are exchanged; and as *participant* articulating opportunities and limits, sometimes representing people who are not at the table, including future generations. This is the role Steve Kelly (1991), the supervisor of the Huron-Manistee National Forest in Michigan, is describing when he says, "I don't *do* anything. I help. I assist." Kelly's typical approach to his forest constituents is: "Here is something that is bothering us. What do *you* think?"

The use of open decision making in communities of interest does not absolve the manager from making tough decisions. But under the best conditions, win-win solutions can emerge from the discussions. There is an incentive to the community of interests. If the community is not able to arrive at a solution, the responsibility for the decision passes by default to the official. Where people prize local control over factors that affect their lives, this should prove a potent stimulus.

Public agencies are gaining experience in collaborative decision processes. In addition to the cases I have discussed, collaborative processes were used in the development of the forest plan for the Bridger-Teton. The limits to the acceptable change process that was pioneered in planning for the Bob Marshall Wilderness are another example of collaborative decision making. On the Huron-Manistee, a group called Friends of the Forest has been working on joint problem solving for more than three years. There are numerous other examples.

To be sure, the community of interests model poses difficult questions of administration, implementation, and responsibility. Here are a few:

- Who is to participate in the community of interests? How do you assure that all relevant, appropriate interests are involved?
- How is the agency or others in the community of interests to represent those not at the table: people around the nation, around the globe, future generations?
- If the group is to resolve issues, who then is accountable?
- What are the dangers of agency manipulation of the process?
- What is the appropriate role of regional offices and the Washington office? Remember, senior agency officials will not have participated in the community of interests dialogue but often are the ultimate decision makers.
- What is reasonable resolution of an issue when there are fundamental value differences that cannot be compromised?

These and other questions remain to be answered. In the meantime, one can count achievements, if not outright successes. In North Carolina, for example, the Forest Service credits relationships built through the open decision-making process for the resolution of an appeal of a timber sale in a roadless area. Forest Planner Pat Cook says the forest planning workshops created a climate in which wilderness advocates and the timber industry could discuss issues in a positive manner. On the Huron-Manistee National Forest, the Friends of the Forest group developed a policy for off-road vehicle use in the forest that has been adopted by the state for lands in Michigan's lower peninsula. On a wider scale, Jeff Sirmon, while Deputy Chief for International Forestry, coalesced a community of interests in the international forestry community to help the agency identify its international role and develop its international forestry programs.

In sum, commonly used approaches to public involvement are not adequate for resolving today's issues in today's decision-making environment. Open decision making in communities of interest is one alternative approach that shows promise. It will require that public land managers *and* interest groups reassess their role in decision making. Managers will have to *share* some decision-making responsibility; members of the community of interests will have to *bear* some of the responsibility for the ultimate decision.

ENDNOTES

1. Senior Fellow, Pinchot Institute for Conservation, and Vice President, Institute for Forest Analysis, Planning, and Policy.

REFERENCES

Blahna, Dale J., and Susan Yonts-Shepard. 1989. Public involvement in resource planning: Toward bridging the gap between policy and implementation. *Society and Natural Resources* 1:209–227.

Heifetz, Ronald A., and Riley M. Sinder. 1990. Political leadership: Managing the public's problem solving. In *The power of public ideas,* edited by Robert B. Reich, p. 187. Cambridge, Massachusetts: Harvard University Press.

Kelly, Steve. 1991. Huron-Manistee National Forests, interview, Cadillac, Michigan, December 11, 1991.

Reich, Robert B. 1990. Policy making in a democracy. In *The Power of Public Ideas,* edited by Robert B. Reich, p. 124. Cambridge, Massachusetts: Harvard University Press.

Shands, William E., V. Alaric Sample, and Dennis Le Master. 1990. *National forest planning: Searching for a common vision,* Washington, D.C.: USDA Forest Service.

Sirmon, Jeff. 1991. Evolving concept of leadership: Towards a sustainable future in forest perspectives. *Journal of Forestry* 1(2): 6–8.

Thomas, Jack Ward. 1994. Statement before the Committee on Natural Resources, United States House of Representatives concerning ''New Directions for the Forest Service,'' February 3, 1994.

Wondolleck, Julia M. 1991. Natural resource management in the 1990s and beyond: Problems and opportunities. Prepared for the Pinchot Institute's Project on Leadership in Natural Resources, December 1991.

Adaptive Tools and Techniques

Overview

The chapters in Part III describe four tested means by which the adaptive visions of natural resource agencies may become reality. Each of these techniques has evolved to meet specific management needs in balancing human wishes and expectations for certain services within legal and ecosystem constraints. Though all of these chapters deal with amenity services by the U.S. National Park Service, the U.S. Forest Service, and Parks Canada, they represent potential opportunities for application of a wide array of ecosystem management issues. Each of these techniques is dedicated to better serving the public, each involves the public in the decision process, and each recognizes limits to supply-side increases and permits more equitable constraints upon demand.

These techniques do not produce absolute solutions, as might be expected from research in physics or microbiology. These are planning tools to help guide, monitor, and evaluate the judgment calls of the manager. Consequently, the rationale for these judgments must be clear. There are few better guidelines than Stankey's for all forms of ecosystem management, regardless as to whether the data are biophysical or social. The data will not program the "correct" answer but will simply assist the judgment. And that judgment is ultimately the responsibility of the manager.

These chapters remind us that there are tools and techniques with thirty or more years of testing that have much value to those who wish to include human behavior within the ecosystem model. The tools and techniques described in this section are but one small subset of the many available for better understanding human behavior. The social science measurement options for ecosystem managers are likely to be as extensive and intensive as those available for biophysical domains. We have given attention to those techniques directed at serving amenity benefits because they are likely to have the most immediate utility in dealing with the broader ecosystem matters of biodiversity.

William R. Burch Jr.

Chapter 9

After 10 and 50: The Adoption and Diffusion of the Visitor Services Project[1]

Gary E. Machlis

The Visitor Services Project (VSP) is a sustained program of visitor studies conducted by the National Park Service. After ten years and fifty visitor studies, it may offer insights into the design, management, and application of a long-term recreation research program. The purpose of this chapter is to assess the project and discuss the implications for other recreation research programs.

First, the project's history is briefly described. The history illustrates the importance of social, institutional, and political factors in the establishment and evolution of applied research programs. Second, the theoretical and methodological approaches used by the VSP are examined, including the resulting trade-offs and limitations. Third, the status of the project is outlined, and examples from recent studies (from Haleakala National Park to the White House) are briefly described. Fourth, the application of these research results (or lack thereof) by managers is evaluated. Fifth, the overall project is assessed as to its strengths, weaknesses, and future potential. Finally, the implication of the project's history to the development of other sustained recreation research programs is discussed.

Applied recreation research can no longer claim to be a young and emergent field. Thirty years have elapsed since the landmark Outdoor Recreation Resources Review Commission (ORRRC) produced its work, and a sizable literature (including concepts, theory, and empirical studies) has emerged and occasionally been synthesized (see, for example, Manning, 1987; Vining, 1990). There are now a large number of social scientists engaged in the recreation research enterprise, and the diversity of theoretical approaches, methodologies, and results suggests an established field of inquiry now searching for a role in management and the setting of social policy.

That role has not markedly grown since the ORRRC report and may have actually declined since the 1970s (contrary to its practitioner's claims). The provision of "usable knowledge" (Lindblom and Cohen, 1979) and its application by recreation managers in North America have increased, though moderately (Machlis, 1991). Concepts like social carrying capacity (Shelby and Heberlein, 1986), methods such as the limits of acceptable change technique, and applications to management concerns about visitor impacts (Graefe et al., 1990) or interpretation (Machlis and Field, 1992) have been accepted and sporadically used by decision makers.

Yet there is little agreement as to how such technology transfer might be organized, conducted, or evaluated. In a survey of 143 recreation researchers, Burton and Jackson (1989) found that 62 percent characterized the field as fragmented and inconsistent. Some recreation agencies actively seek and support recreation research; others do not. The level of support varies dramatically between agencies, within administrative regions, and from year to year. There is a gap between managers' "need-to-know" and the information and expertise available to them (Jordan et al., 1992). Agencies like the National Park Service have been admonished by various authors and study groups for a lack of social science research capability (Meis, 1990; National Parks and Conservation Association, 1989; National Park Service (NPS), 1992).

This is partly a problem in the organization and delivery of science; new approaches may be required. One strategic approach is the development of *sustained* research programs, which have been described as

> . . . a series of interrelated studies, each of which is associated with an identifiable stage of development where the stages are ordered according to their capacity to identify and solve sociological problems (Cohen, 1989, 292).

A sustained recreation research program has the following characteristics: (1) it involves a series of interrelated applied recreation research studies, (2) its focus is on providing usable knowledge for recreation managers, and (3) it has an organizational and institutional identity. The development of sustained recreation research programs may be one way that the gap between recreation managers' needs and social science research can be narrowed. The relative lack of such programs may be one reason this gap exists.

Hence, a crucial question is, Under what conditions will sustained recreation research programs succeed? This is a problem in the adoption and diffusion of an innovation. Treated this way, it is amenable to solution through the use of diffusion and innovation theory and findings.

The purpose of this chapter is to examine one sustained recreation research program, the Visitor Services Project (VSP) of the U.S. National Park Service, and to search for lessons as to how recreation research might be carried out in the 1990s. First, the VSP is briefly described, including its development, scope, methods, and results. Second, Rogers' (1983) theory of the adoption and diffusion of innovations is described, and current critique and application of the theory are noted. Third, the VSP is evaluated in light of the insights that emerge from understanding the processes of innovation.[2] Finally, the implications for other sustained recreation research programs are discussed.

THE VISITOR SERVICES PROJECT

The VSP began in 1982 with a single pilot study conducted by the Cooperative Park Studies Unit (CPSU) at the University of Idaho. The CPSU is a research station of the National Park Service (NPS) and had previously been involved in a variety of visitor studies—each one carried out separately for an individual client (usually a park unit). The pilot study was conducted at Grand Teton National Park, and its purpose was to develop a technique for mapping (or inventorying) interpretive services, one that could be applied at various parks. Evaluation by managers was both harsh and encouraging: the information was valuable, especially the information about visitors, but its delivery was inadequate.

The project was redesigned, this time treating park managers as potential adopters, using visitor data as an innovation, and having an objective of a sustained recreation research program providing usable knowledge. Further pilot studies were conducted at Yellowstone (1983), Glacier (1984), and Crater Lake (1985). In each case, park staff provided detailed evaluations, and improvements in methodology, cost effectiveness, and delivery were attempted.

The VSP was made operational in 1985, with the results of its visitor studies now treated as "usable knowledge" for decision making. The essential elements of the methodology and delivery system were in place:

1　a "client park" would request a VSP study;
2　a mail-back questionnaire would be designed in close consultation with park staff yet conforming to a standardized format;
3　using the questionnaire and a brief front-end interview, CPSU staff would conduct an intensive one-week sampling of visitors;
4　the data would be analyzed and displayed in graphic and map form; and
5　a workshop would be conducted at the park on how to use the data in solving park problems. Figure 1 outlines the process.

From 1985 to 1988 the VSP was supported by the NPS on a year-to-year basis. By 1988, the demand for visitor studies exceeded the capacity of the CPSU to conduct them. The director of the NPS and its ten regional directors decided to more formally adopt the VSP. The project was expanded and formally institutionalized within the agency. NPS employees were assigned to the CPSU

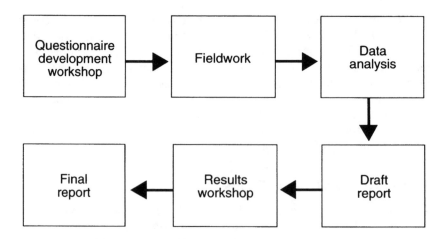

Figure 1   Steps in a typical visitor services project study.

to assist in the visitor studies, financial resources were increased, an advisory committee was established (with charter) to select parks for study, and the number of studies to be conducted each year was increased significantly.

Awareness of the project grew, both in Washington, D.C., and in the field. By 1988, a survey of 713 NPS interpretation managers and employees revealed that 62 percent of all respondents knew about the VSP (Raithel, 1988). The survey showed that higher-graded employees were more likely to know about the VSP than lower-graded employees (see Figure 2). Over half of interpretation employees in all types of parks (natural, historical, and recreational) knew about the VSP, although knowledge of the VSP was more frequent among employees in natural areas (over 70 percent).

In 1991, studies were conducted at six parks, ranging from the Joshua Tree National Monument to the White House. Ten parks were slated for VSP studies in 1992. The methodology has continued to improve; for example, response rates have increased from an average of 37 percent in the early studies to 83 percent for the 1991 studies. The technique has been adopted internationally in various locations including the Galápagos Islands, Mount Kilimanjaro in Tanzania, and the Caribbean island of Saint Kitts. The program has expanded to include the development of evaluation techniques for interpreters (see Machlis and McKendry, 1991; Medlin and Machlis, 1991; Medlin, 1992). Several descriptions of the project have been prepared (see Littlejohn and Machlis, 1989, 1990; Dolsen, 1990; Dolsen and Machlis, 1991). After ten years and fifty studies, the VSP is an ongoing and sustained research program.

THE ADOPTION AND DIFFUSION OF INNOVATIONS

The study of adoption and diffusion has a long tradition in sociology, and case studies have included agricultural innovations, acceptance of new pharmaceuticals by doctors, and new industrial processes. In 1962, E. M. Rogers published

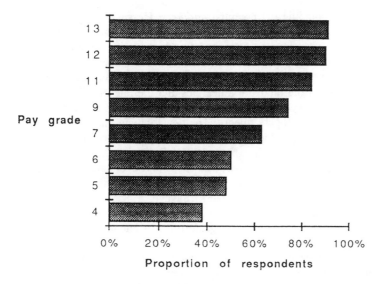

Figure 2 Knowledge of the Visitor Services Project by grade.

Diffusion of Innovations, and this work and its more recent edition (Rogers, 1983) provide a useful framework. Rogers defines *diffusion* as

> the process by which an innovation is communicated through certain channels over time among members of a social system (Rogers, 1983, 5).

To Rogers, an *innovation* is an idea, practice, or object that is perceived as new by an individual or other unit of adoption. There are several characteristics of an innovation that will significantly influence its adoption. These include

Relative advantage: the degree to which an innovation is perceived as better than the idea that it supersedes

Compatibility: the degree to which an innovation is perceived as being consistent with the existing values, past experiences, and needs of potential adopters

Complexity: the degree to which an innovation is perceived as difficult to understand and use

Trialability: the degree to which an innovation may be experimented with on a trial basis

Observability: the degree to which the results of an innovation are visible to others

These characteristics of the innovation, as perceived by the potential adopters, help explain the rate of adoption for specific innovations.

Crucial in the adoption process are characteristics of the *change agent,* i.e., those individuals or organizational units that attempt to influence adoption decisions. Rogers emphasizes the need for change agents to have social interaction with (and an understanding of) potential *adopters.* He suggests that there are different classes of adopters—from "innovators" to "laggards"—and each class may have a different set of reasons to adopt an innovation. Figure 3 shows that typical diffusion patterns follow an "S-curve," with change agents attempting to influence innovators and opinion leaders, and later adopters joining in as the uncertainty of the innovation declines and the rate of adoption slows.

Current literature suggests that innovation diffusion is more complex than the process described by Rogers' theory. The development and diffusion of an innovation may not be separable processes; an innovation is rarely fully developed when diffusion begins (Silverberg, 1991). Rogers underemphasized the importance of spatial relationships (Brown, 1981) and overstated the effectiveness of external change agents (Stewart, 1982). There are contradictory findings about the characteristics of early adopters (Savage, 1985). Nevertheless, the general argument of how innovations are adopted and diffused is both widely accepted and useful in designing strategic efforts to encourage the adoption process.

Several studies have examined how applied social science is adopted by organizations and diffused within and among them. In a general analysis, Weiss (1980) found three barriers to diffusion: (1) the research-producing system may be deficient, producing irrelevant results; (2) the information-transmission system may be ill-suited for communicating with clients; and (3) decision makers

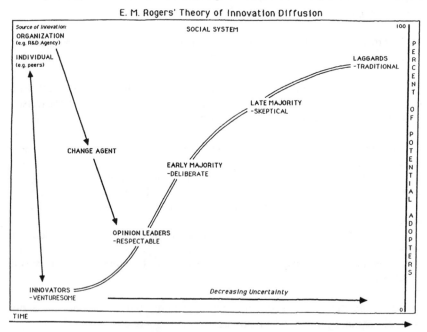

Figure 3 Typical diffusion pattern. Adopted with the permission of The Free Press, a division of Simon and Schuster, from *Diffusion of Innovations,* 4th ed., by E. M. Rogers. Copyright © 1995 by E. M. Rogers. Copyright © 1962, 1971, 1983 by The Free Press.

may be unwilling or unable to use social science research. Hulme (1990), using several case studies, found that indigenous knowledge of clients can be incorporated into the social science research process, i.e., the innovation itself. Savage (1985) found that timeliness is an important characteristic of a social science innovation; Johnson and Field (n.d.) note that the work cycles of universities and most clients (including recreation agencies) are not the same.

In his book *Effective Social Science,* Barber (1987) studied eight cases of applied social science programs in economics, political science, and sociology. He found several patterns in how applied social science programs become effective and are diffused within organizations. Most effective programs were "quiet cases," where the effectiveness was recognized by specific clients and subspecialists within the social sciences. Effective social science programs resembled natural science programs (such as several biomedical programs). The active role of Washington, D.C., in science policy and organizational management made contact with the nation's capital a critical responsibility. Small-scale research organizations, with identity and institutional stability beyond the individual scientists, were essential steps in increasing effectiveness. The influence of sponsors was continual, and effective programs set up advisory panels to buffer the research from outside interference, balance diverse beliefs and interests, and gain support from funding agencies.

THE VISITOR SERVICES PROJECT AS AN ADOPTED INNOVATION

The Visitor Services Project can be considered an adopted innovation within NPS. It began in 1982 as a single pilot study peripheral to NPS managers' interests. By 1992, it was institutionalized, expanded into a national program, and visible within the agency and the community of recreation researchers. What components of the theory of adoption and diffusion might explain this growth?

The VSP was explicitly conceptualized as an innovation for recreation managers: a visitor study *package* that included design, implementation, and follow-up support. It was not a scientific innovation, as it relied on existing methodology, standard analysis techniques, and reports generated with off-the-shelf graphic software. It was *perceived* as an innovation by managers because it offered a new delivery system for needed visitor information.

The unit of adoption was a critical choice; what is appropriate in one social system or organization may be ineffectual in another. Decision making in the NPS is relatively decentralized, with superintendents of individual park units having significant authority and information needs. Hence, the potential adopters were park units and their superintendents, and the VSP was designed to maximize effectiveness at this level of the organization. As the project expanded, regional offices of the agency (and regional directors) became crucial clients, and eventually the Washington, D.C., office of the NPS and the director of the NPS (there have been three different incumbents during the years 1982–1992)

CASE STUDY: THOUGHTS ON A VISITOR SERVICES PROJECT AT BRYCE CANYON NATIONAL PARK

Bob Reynolds

The Second Canada/US Workshop on Visitor Management in Parks, Forests and Protected Areas, May 15, 1992

I arrived in Bryce Canyon National Park as Superintendent in June of 1988. The park had a new General Management Plan, completed just two years earlier. Visitation had been increasing rapidly, and the capabilities of the park to deal with the interpretive and informational needs of the park visitors were being severely taxed. At the same time, the park was poised to receive a significant infusion of funding to implement a recently completed Interpretive Prospectus, and there was a possibility for increases in staffing. We recognized the possibility of reorganizing to better meet visitor needs.

Most of the knowledge of park visitors, and of the values that park visitors placed on our interpretive and informational services, was anecdotal in nature or derived from studies that were only peripherally about the kind of visitor characteristics we were most interested in. We didn't know if they thought an evening program or a horse ride concession were worth anything, and we didn't really know much else about them or their desires.

The visitor survey was conducted about a month after my arrival during one week in July 1988 by Gary Machlis of the Cooperative Park Studies Unit at the University of Idaho, as part of the National Park Service's Visitor Services Project (VSP). Visitors were contacted at the park entrance station. Front-end interviews were conducted with a sample of randomly selected visitors. The visitors being interviewed were given a questionnaire to be completed during their trip and returned by mail. The questionnaire asked visitors where they went, what they did, how much money they spent in the area and when they would prefer to attend conducted activities. Visitors were also asked to rate the importance and quality of certain services or facilities. Additional space was provided for comments. A total of 484 questionnaires were distributed and there was an impressive 84% return rate.

Little in the report was a surprise. Most of it confirmed what we already thought we knew. But it was presented differently, and it did lead us to think differently about the information. Several operational changes resulted. For instance, most visitors were visiting for the first time, but had already visited one of the nearby national parks. The orientation information we had been providing at the visitor center, the entrance station and in the park newspaper was aimed at a first-time visitor, one without any national park experience. Since we could assume most visitors had been to one of the other near-by parks, we could build on that experience—not only in orienting them to the park, but in our interpretive program generally. From that insight came an analysis of the pre-arrival orientation materials. We developed a new pre-visit informational publication to send to people calling or writing for information. That same publication was also provided to area information centers run by other organizations. It provided an impetus for us to increase our work with other state and federal agencies on joint information services.

We also learned from the VSP that a third of our visitors were foreign, and most of those were European. Survey comments by foreign visitors pointed out the relative paucity of foreign language materials, and absence of interpretive programming. That insight led to an effort to translate more publications into German and French. And we began looking for seasonal interpreters with language skills.

We found that many of our interpretive programs were scheduled at inconvenient times for visitors, though they were convenient for scheduling park rangers. The next summer, walks and talks were scheduled at the times identified as most preferable, and attendance increased—in some cases to the point of overload.

Bryce Canyon is near the western edge of a time zone, so traditional summer evening programs were started as late as 10 or 10:30 to have enough darkness. In 1989, the auditorium at the visitor center was remodeled allowing us to use it for earlier evening programs. They proved so popular that eventually they became a regular addition to the interpretive menu. They were particularly popular with families who told us that the regular programs were too late for kids.

We also learned from the survey that two-thirds of the visitors never attended guided walks and evening programs. If not, what were they doing? And how could we modify our approach to ensure that most visitors received the park story, safety messages, and valuable resource information? Respondents told us resoundingly that visitor center information, printed materials, and directional signing was very or extremely important. Most of them stopped at the visitor center, enjoyed scenic driving but stopped at viewpoints, and two-thirds went for a hike.

We took the information about what they considered important, combined with an analysis of their activities, and asked ourselves if we were putting our fiscal and personnel resources in the most effective places. We decided to make some changes. Visitor center staffing was increased and hours expanded, and the park nature center that had been closed for many years was reopened as a second information station within the park. Wayside exhibits were moved up in the priority list, with special emphasis on orientation posters to be placed in front of the visitor center. Two seasonal interpretive positions were combined to fund one year-round interpretive specialist with particular expertise in graphics, writing and publication production. We placed greater emphasis on non-personal interpretive programming. We produced several new site bulletins, redesigned the park newspaper, and worked closely with commercial publishers and the cooperating association to produce new interpretive materials. The Bryce Canyon Natural History Association hired a new publications coordinator so that they could be more professional and proactive in the publications program.

None of the changes were stimulated by the Visitor Service Project alone. The information derived from the project was often the stimulus to begin the questioning and analysis that led eventually to a change, or was a part of the data used in the decision-making process.

When the analysis was completed, we arranged for Dr. Machlis to give a seminar to the business community. The economic data also proved useful to me as a park manager in developing a very rough model of the impact of the park on the local economy, which I used in discussions with local officials and decision-makers.

The visitor survey was a snapshot. We felt it was accurate for the main visitor season, but only then. We used the information for decision-making, but often had to adjust it based on limited data from other seasons, or our gut feeling of how other times varied from the week surveyed. Unfortunately, the data became dated very quickly. Within just a couple of years, our personal observations and some other park data were indicating that visitor patterns were changing. We wanted in the worst way to repeat the survey to confirm the trends we thought we saw emerging.

From our experience with the VSP come a few suggestions. We should encourage parks to plan for multi-season surveys right from the start. And we should encourage follow-up surveys so parks can have up-to date information.

So was the VSP worth the effort? Yes, it certainly was. Most of the real work was done by the folks from Idaho. But the value of any survey is to a degree dependent upon the willingness of park staff to put good thought into its design and implementation phases. At Bryce Canyon, that was Susan Colclazer, the Chief Interpreter, and her assistant, Margaret Littlejohn. The data were particularly valuable because they were so specific and in-depth. And the data relating to values placed on park activities and functions, along with information about how they chose to make use of their time were very useful. The VSP was one of the best investments we made.

were involved in adoption decisions. However, the focus of the VSP has remained on serving managers at individual parks.

Table 1 illustrates key features that were included in the VSP to address Rogers' suggested characteristics of a successful innovation. Because of the conservatism within NPS organizational culture (Foresta, 1984) and the budget limitations that give individual superintendents little opportunity for innovation, the characteristics of relative advantage and compatibility were stressed in the

Table 1 Visitor Services Project 1982–1992

Development studies			Completed studies			
1982–1986	1987	1988	1989	1990	1991	1992
Grand Teton NP	Gettysburg NMP	Craters of the Moon NM	Everglades NP	Canyonlands NP	Jean Lafitte NHP	Big Band NP
VSP evaluation	Independence NHP (summer)	Bryce Canyon NP (summer)	Statue of Liberty NM	White Sands NM	Joshua Tree NM	Frederick Douglass Home NHS
Yellowstone NP/ Mount Rushmore NM	Valley Forge NHP	Glen Canyon NRA	White House Tours (summer)	National Monuments	White House Tours (spring)	GWMP-Glen Echo Park
Yellowstone NP	Harpers Ferry NHP	Denali NP	Delaware Water Gap NRA	Kenai Fjords NP	Naatchez Trace Parkway	Bent's Old Fort NHS
North Cascades NP	Grand Teton NP (summer)		Lincoln Home NHS	Gateway NRA	Stehekin-North Cascades NP	Jefferson National Expansion Memorial
Crater Lake NP	Yellowstone NP (summer)		Yellowstone NP (summer)	Petersburg NB	City of Rocks NR	Zion NP
	Mesa Verde NP		Muir Woods NM	Death Valley NM (summer)	White House Tours (fall)	New River Gorge NR
	Independence NHP (spring/ fall)			Galacier NP		Klondike Gold Rush NHP
	Colonial NHP (summer/fall)			Scott's Bluff NM		Arlington House
	Shenandoah NP (summer/fall)			John Day Fossil Beds NM		

NP, national park; NM, national monument; NHP, national historical park; NRA, national recreation area; NHS, national historic site; NB, national battlefield; GWMP, George Washington Memorial Parkway

design of the VSP technique. Because that same organizational culture encour-
ages close interaction between superintendents (and an awareness of opinion
leaders and their influence), observability was stressed in the diffusion strategy.

The strategic approach for diffusing the VSP within the NPS was explicitly
organized around Rogers' adoption curve, described in Figure 3. Innovators and
opinion leaders were identified, and pilot studies were conducted at parks that
served to demonstrate the adoption potential to the general superintendent corps.
Various communication channels were used to disseminate information about
the project, including training programs for new superintendents (future adopters), a newsletter describing current activities, and a short videotape that could
be sent to all potential adopters. Evaluations of completed studies by early-
adopting superintendents were shared with late adopters; these served as valuable
"testimonies." The establishment of an advisory board made up of NPS decision
makers at park, regional, and national levels served to establish the legitimacy
of the program, ensure its continuing response to client needs, and increase the
number of communication channels available.

The VSP techniques often had to be revised, as new requirements (such as
stringent Office of Management and Budget rules) were put in place. Since the
organizational environment was continually changing, the innovation had to
evolve, rather than be treated as a static entity. This was time-consuming but
encouraged technical improvements as well. Close interaction with superintendents and other field managers, including thorough evaluation of the pilot studies,
allowed the delivery system to be designed and modified around the potential
adopter's requirements. Evaluation by early adopters provided for additional im-
provement.

Setbacks inevitably occur in the diffusion of an innovation, and the VSP is
no exception. In many cases, these setbacks were the result of ignoring some
aspect of the adoption process. For example, VSP staff underestimated the threat
the project represented to established programs (such as other CPSUs, consulting
firms, and professional staff at the regional and national offices of the agency).
Several individuals and units perceived increased competition or even substitu-
tion; efforts to cooperate with or co-opt these threatened interests were too little
and too late. Also, NPS management continually changes, and with each change,
the process of dealing with new potential adopters had to begin again. This
occurred at all levels of the organization: a visitor study might be initiated under
one superintendent, only to be completed for another; a supportive regional
director might be replaced with one unaware of the VSP. A continual program
of "adopter education" was required and yet only sporadically conducted. Both
these deficiencies slowed the adoption and diffusion of the VSP and reduced
the effectiveness of the program.

CONCLUSION: IMPLICATIONS FOR SUSTAINED
RECREATION RESEARCH PROGRAMS

These results have potential implications for the design of sustained recreation
research programs. First, the successful adoption of the VSP by the NPS suggests

that this kind of applied science delivery system may be a viable approach for other programs and in other recreation agencies. Sustained research programs are an important alternative to the fragmented, individualistic approach that characterizes much of current recreation research.

Second, the explicit use of adoption theory significantly influenced the VSP diffusion process; it may be relevant in other, similar situations. For example, explicitly reconceptualizing the products of recreation research as innovations and resource managers as potential adopters may yield strategic benefits. There is some irony in suggesting that social scientists should pay attention to social theory in the design of their research programs. Yet technology transfer is not well taught in graduate school, the literature on recreation research does not include research on program design or implementation, conferences and symposia largely ignore the topic, and recreation agencies are reluctant to fund innovative research and development efforts.

Finally, general principles derived from adoption theory might serve as practical guides in designing and administering recreation research programs. These principles will need to be carefully generated and must be adapted to specific situations. Various recreation research programs may represent radically different innovations, organizational cultures may create dramatically different "adoption environments," and the units of adoption, change agents, and opinion leaders may vary case-to-case and over time. Hence, there is no one best way to organize an adoption and diffusion strategy, nor is there a single model for the conduct and delivery of recreation research. Those that would manage the social sciences within recreation agencies would do well to heed that insight.

ENDNOTES

1. This keynote speech was presented at the Second Canada-U.S. Workshop on Visitor Management in Parks, Forests, and Protected Areas, May 19, 1992, Madison, Wisconsin. The paper is adapted from the following:

Machlis, Gary E., and M. Jeannie Harvey. 1993. The adoption and diffusion of recreation research programs: A cast study of the Visitor Services Project. *Journal of Park and Recreation Administration* 11(1):49–65.

2. The scope of this analysis is largely limited to the adoption and diffusion of the VSP. Its usefulness (as measured by managers' use of VSP results), efficiency (relative cost), and methodological issues are treated elsewhere (Littlejohn and Machlis, 1990; Dolsen and Machlis, 1991).

REFERENCES

Barber, Bernard. 1987. *Effective social science: Eight cases in economics, political science, and sociology.* New York: Russell Sage Foundation.

Brown, Lawrence A. 1981. *Innovation diffusion: A new perspective.* New York: Methuen.

Burton, Thomas L., and Edgar L. Jackson. 1989. Leisure research and the social sciences: An exploratory study of active researchers. *Leisure Studies* 8:263–280.

Cohen, Bernard. 1989. *Developing sociological knowledge: Theory and method,* 2nd ed. Chicago, Illinois: Nelson Hall.

Dolsen, Dana. 1990. Surveys of national park experiences by the Visitor Services Project. *Park Science* 10(4): 5–6.

Dolsen, Dana E., and Gary E. Machlis. 1991. Response rates and mail recreation survey results: How much is enough? *Journal of Leisure Research* 23(3): 272–277.

Foresta, Ronald A. 1984. *America's national parks and their keepers.* Washington, D.C.: Resources for the Future.

Graefe, A. R., F. R. Kuss, and J. J. Vaske. 1990. *Visitor impact management: The planning framework.* Washington, D.C.: National Parks and Conservation Association.

Hulme, David. 1990. Agricultural technology development, agricultural extension and applied social research. *Sociologia Ruralis* XXX(3/4).

Johnson, Darryll R., and Donald R. Field. n.d. Toward a practical perspective: A preface to applied sociology. Seattle: University of Washington, Cooperative Park Studies Unit.

Jordan, Charles R., Gary E. Machlis, Derrick A. Crandall, Diane H. Dayson, John P. Debo Jr., Robert C. Herbst, Laura A. Loomis, and John Reynolds. 1992. Final report: Working group on park use and enjoyment. In *National parks for the 21st century: The Vail agenda.* Denver, Colorado: National Park Service, Denver Service Center.

Lindblom, Charles E., and David K. Cohen. 1979. *Usable knowledge: Social science and social problem solving.* New Haven, Connecticut: Yale University Press.

Littlejohn, Margaret, and Gary E. Machlis. 1989. The Visitor Services Project and beyond. *Interpretation* (Summer).

Littlejohn, Margaret, and Gary E. Machlis. 1990. A diversity of visitors: A report on visitors to the National Park System. Moscow: University of Idaho, Cooperative Park Studies Unit.

Machlis, Gary E. 1991. Social science and protected area management: The principles of partnership. Plenary paper presented at the World Parks Congress, Caracas, Venezuela, February 10–21.

Machlis, Gary E., and Donald R. Field. 1992. *On interpretation: Sociology for interpreters of natural and cultural history,* rev. ed. Corvallis: Oregon State University Press.

Machlis, Gary E., and Jean E. McKendry. 1991. A self-critique of the interpretive program at Whitman Mission National Historic Site. Moscow: University of Idaho, Cooperative Park Studies Unit.

Manning, R. E. 1987. *Studies in outdoor recreation: A review and synthesis of the social science research in outdoor recreation.* Corvallis: Oregon State University Press.

Medlin, Nancy. 1992. A self-critique of the interpretive program at Nez Perce National Historical Park. Moscow: University of Idaho, Cooperative Park Studies Unit.

Medlin, Nancy, and Gary E. Machlis. 1991. Focus groups: A tool for evaluating interpretive services. Moscow: University of Idaho, Cooperative Park Studies Unit.

Meis, Scott M. 1990. The socio-economic function of the Canadian Parks Service as a model for the U.S. National Park Service and other agencies: An organizational framework for managing natural resource recreation research. In *Social science and natural resource recreation management,* edited by Joanne Vining, pp. 33–43. Boulder, Colorado: Westview Press.

National Park Service. 1992. *National parks for the 21st century: The Vail agenda.* Washington, D.C.: National Park Service, Denver Service Center.

National Parks and Conservation Association. 1989. *Interpretation: Key to the park experience,* vol. 4. Washington, D.C.: National Parks and Conservation Association.

Raithel, Kenneth, Jr. 1988. Interpreters survey report. Washington, D.C.: National Park Service.

Rogers, Everett M. 1983. *Diffusion of innovations,* 3rd ed. New York: The Free Press.

Savage, Robert L. 1985. Diffusion research traditions and the spread of policy innovations in a federal system. *The Journal of Federalism* 15(Fall).

Shelby, Bo, and Thomas A. Heberlein. 1986. *Carrying capacity in recreation settings.* Corvallis: Oregon State University Press.

Silverberg, Gerald. 1991. Adoption and diffusion of technology as a collective evolutionary process. *Technological Forecasting and Social Change* 39:67–80.

Stewart, David W. 1982. The diffusion of innovations: A review of research and theory with implications for computer technology. Presented at the Annual Convention of the American Psychological Association, Washington, D.C., August.

Vining, Joanne, ed. 1990. *Social science and natural resource recreation management.* Boulder, Colorado: Westview Press.

Weiss, Carol. 1980. *Social science research and decision making.* New York: Columbia University Press.

Chapter 10

Benefits-Based Management: A New Paradigm for Managing Amenity Resources[1]

Martha E. Lee

B. L. Driver

During the past decade, we have witnessed a change in focus among public land management agencies. Whereas recreation and other amenity resources such as wilderness, visual, and heritage resources were formerly either taken for granted or perceived as trivial or superfluous (Driver et al., 1986), public amenity resource managers are beginning to focus more explicitly on the goods and services they are providing. This change has occurred in part because recreation managers and policy makers have been forced to more clearly justify their budget requests in these times of scarce agency resources. Increased environmental awareness and public involvement in decision-making activities have also contributed to the need for managers to better understand and articulate the human values associated with recreation and other amenity resources.

A research and management framework has emerged to help land managers and policy makers more clearly define amenity resource management outputs. This framework, called benefits-based management (BBM), seeks to understand and empirically document the benefits people receive from amenity resources, including participation in recreation and leisure activities. In other words, BBM addresses the question, What "good" do amenity resources, including leisure and recreation, do for people?

Benefits can be physical, social, and psychological. Benefits are realized by individuals and groups of individuals (e.g., families, communities, and society at large). Potential benefits from leisure and recreation, for example, include the spiritual benefits people might gain from exposure to cultural resources or wilderness, the pollution-reducing and scenic benefits of trees in an urban environment, and the feelings of stewardship realized from protecting and preserving wildlife. BBM is based on the premise that before management agencies can truly serve and meet the needs of people (both users and nonusers of public lands), they must understand what people want and what managers can and are providing, articulate those wants, and develop and deliver benefit-related outputs.

Since the early 1980s, the body of scientifically documented evidence about the benefits of leisure and other amenity resources has grown rapidly. A wide variety of benefits are being identified and specified more clearly, their scope and magnitude are being quantified, and their values to individuals are being established more accurately. BBM is becoming more widely accepted by public land managers. For example, the U.S. Forest Service recently adopted and will incrementally implement BBM, and the concept is now being taught at a U.S. Forest Service sponsored recreation short course at Clemson University, which is attended by managers from across the United States. Other federal, state, and municipal agencies and departments are currently involved in pilot projects to assess the usefulness of the BBM approach. The purpose of this chapter is to describe BBM, its origins, how it differs from other recreation management approaches, and its current status, beginning with a brief discussion of benefits and their relation to recreation activities and experiences.

WHAT IS A BENEFIT?

Most simply, a benefit is defined as an improved condition (a gain) or the prevention of a worse condition (usually through some maintenance function) to an individual or group of individuals (e.g., a family, community, or society) (Driver and Peterson, 1987). For example, a benefit occurs if one's health is either improved or maintained. A healthy person may or may not realize a positive change in health from continued jogging, but if he or she stopped jogging, there could be a decline in health.

Figure 1 illustrates how benefits are realized, beginning with engagement in a recreation activity, and shows the relationship between recreation or leisure activities and benefits. It is important to recognize that while engaging in recreation or leisure activities causes positive changes, it also serves many maintenance functions we may not consciously recognize that prevent negative changes or a worse condition, such as a decline in physical health.

Other benefits or improvements we can recognize or feel are change in blood pressure, a more positive mood, or stronger feelings toward a loved one. Some benefits occur during a leisure or recreation engagement (while viewing a scenic vista or while watching a program on public television, for example), others occur immediately after, and still others some time later. Thus, some benefits,

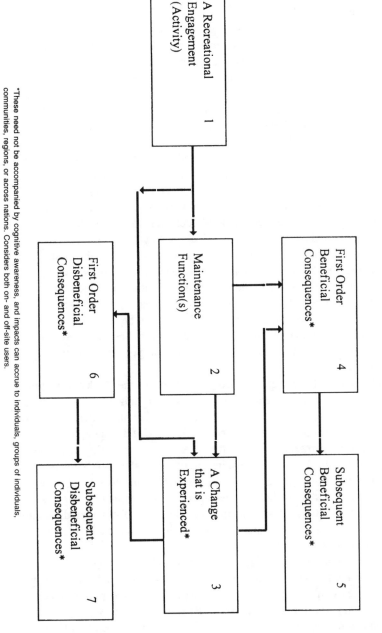

RELATIONSHIP BETWEEN RECREATION ACTIVITIES AND BENEFITS

A Recreational Engagement (Activity) 1		

Maintenance Function(s) 2

First Order Beneficial Consequences* 4

First Order Disbeneficial Consequences* 6

A Change that is Experienced* 3

Subsequent Beneficial Consequences* 5

Subsequent Disbeneficial Consequences* 7

*These need not be accompanied by cognitive awareness, and impacts can accrue to individuals, groups of individuals, communities, regions, or across nations. Considers both on- and off-site users.

Figure 1 Relationship between recreation activities and benefits.

a positive change in mood, for example, might be of short duration. Others might be long lasting. Also, many benefits realized at one point contribute to other additional benefits realized later.

Some examples of immediate and subsequent benefits of leisure, recreation, and other amenity resources may include reduced blood pressure, enhanced self-image, relief from everyday life stresses, strengthened family cohesion, increased work productivity, increased understanding of natural processes, a stronger environmental ethic, greater knowledge and pride in a nation or heritage, and a greater sense of stability and balance or increased nurturing among social groups. Research has shown that the range of benefits can be extensive, including physiological benefits (Ulrich et al., 1990), improved states of mind (Driver et al., 1987b), learning (Roggenbuck et al., 1990), and spirituality (McDonald and Schreyer, 1991). It could include benefits to families (Orthner and Mancini, 1991), organizations (Ellis and Richardson, 1991), communities (Allen, 1990), and societies (Driver and Brown, 1987). It also includes nonhuman benefits, such as those derived from preserving natural ecosystems (Rolston, 1985).

Social scientists at the U.S. Forest Service's Rocky Mountain Forest and Range Experiment Station (RMS) have taken a leading role in developing a better understanding and knowledge base of the benefits of amenity goods and services, particularly the leisure and recreation opportunities provided on public lands. They have facilitated the coming together of a diverse group of scientists, researchers, and resource managers to address the issue of benefits, specifically, to (1) identify what is known about benefits, and (2) to define how knowledge about benefits can be used by public makers and managers of recreation and amenity resources.

In May of 1989, B. L. Driver and George Peterson of the Rocky Mountain Station and Perry Brown of Oregon State University organized a workshop at Snowbird, Utah, at which fifty-seven experts from six countries met to assess the state of knowledge about the many benefits of leisure and amenity resources and to give direction to research. Out of this conference came a book, *Benefits of Leisure,* published by Venture Publishing in 1991. Of the book's thirty-five chapters, the first five document the needs for research on the benefits of leisure, the next twenty-one are state-of-knowledge papers about specific types or classes of benefits, eight record the professional opinions of experts (from eight different disciplines) about problems associated with research on the benefits of leisure, and there is an integrative summary chapter.

The Snowbird state-of-knowledge workshop and the resulting text created considerable demand to translate how information about the benefits of amenity goods and services could be used by public natural resource policy makers, planners, and managers. In response, Driver, Peterson, and Brown organized a follow-up applications workshop in Estes Park, Colorado, in May 1991. That workshop was attended by seventy people, about equally divided between researchers/academics and amenity resource practitioners, representing all levels of government and the private sector in Canada and the United States. The result was the preliminary development of the concept of BBM of amenity resources.

BENEFITS-BASED MANAGEMENT

Benefits-based management requires that the managing agency target explicitly stated types of "benefit opportunities" that will be provided at designated sites and areas and then write and implement time-bound management objectives and prescriptions (with guidelines and standards) developed to ensure that these targeted benefit opportunities will be provided. This management approach is somewhat unique compared to prior recreation resource management frameworks that have focused primarily on activities or experiences.

BBM builds upon and is an extension of two recreation resource management frameworks: activity-based management (ABM) and experience-based management (EBM). A comparison of these three approaches is presented in Figure 2. ABM viewed a recreation opportunity as an option for people to participate in a specified activity (camping, fishing, tennis, hiking). This approach was primarily supply oriented, with attention given to the attributes of recreation settings required to produce different types of activities. There was little attention given by managers to visitor satisfaction or what recreationists got from use of the opportunity. Management objectives were defined in terms of numbers of activity opportunities to be provided, with little concern for what constituted a quality recreation opportunity.

EBM built on and supplemented (not replaced) ABM. This approach broadened ABM to offer a more behaviorally oriented definition of a recreation opportunity as a chance to engage in a preferred activity within desired settings to realize desired experiences. In this definition, recreation activities are behaviors such as hunting, hiking, and fishing. Settings are the places where activities take place and include all physical resources (e.g., topography, vegetation, water), social conditions (e.g., numbers and behavior of other people), and managerial conditions (e.g., fee systems, regulations, permits, facilities) of those places.

Experiences are defined as psychological outcomes or specific types of responses, such as feeling relaxed, invigorated, closer to members of one's group or family, more self-reliant/confident, or more knowledgeable about something. EBM focuses attention on what happens to the user: what types of experiences are demanded, which are realized, and of what quality. Within EBM, the concept of product or management output is expanded to include not only the activity opportunity but also the specific types of experience opportunities produced. This approach facilitates a more systematic understanding of the role of recreation setting attributes in creating not only activity opportunities, but also experience opportunities.

The EBM approach was a significant advancement over ABM because managers could now explicitly include the concept of experience opportunity in

(ABM) *Activity-Based Management*	(EBM) *Experience-Based Management*	(BBM) *Benefits-Based Management*
—simplistic, defines recreation only as participation in an activity	—more complex, defines recreation as a psychological state; targets not only activities, but experiences	—considers activities and all types, of experiences, both physiological and psychological
—no attention given to how recreationist is affected or impacted by provision of an activity opportunity	—explicitly relates setting attributes to visitors' experience and activity demands; it is the basis for the ROS	—considers all types of benefits, including those to community and environment, as well as on-site users
—supply-oriented; focuses on facilities or resources; little attention to demand	—consumer-oriented; focuses on desired types of experiences	—consumer-oriented; considers not only immediate, but long-term benefits
—inputs and outputs defined in the same terms—numbers of users	—requires understanding of both supply and demand, e.g., types of experience opportunities available and desired	—requires information from the public(s) served on the types of benefit opportunities they desire
—little consideration for quality, user satisfaction	—more focus on quality, requires analysis and evaluation of user satisfaction	—even more focus on quality, requires a more explicit definition of the product being produced
—management objectives and recreation production process oriented toward activities	—management objectives explicitly target experience opportunities to be provided, when, where, for whom, and in what amount	—management objectives explicitly specify types of benefit opportunities to be provided, when, where, for whom, and in what amount

Figure 2 Comparison of ABM, EBM, and BBM.

management objectives; that is, they could specify types of experience opportunities (for solitude, learning, physical fitness, family togetherness, escape, skill development, etc.) to be targeted as a product of management. Thus some areas/resources are designated to provide one set of experience opportunities, and other areas/resources are designated to provide other experience opportunities. To implement this approach, information is needed about the relationship between particular types of experiences and the activities and settings in which they occur.

EBM forms the basis for the Recreation Opportunity Spectrum (ROS) (Driver et al., 1987a), a recreation resource inventory and management system where explicitly targeted types of experience opportunities are inventoried according to specific setting, activity, and experience-defining criteria. Use of ROS enables managers to explicitly target and manage for specific types of experience opportunities within ROS-defined management zones designated on the ground. ROS is used extensively to manage recreation resources on public lands managed by the U.S. Forest Service and the Bureau of Land Management.

BBM is the logical extension of EBM and is based on the idea that (1) the reason public recreation opportunities are provided is because people benefit from them, and (2) management will be most responsive, efficient, and effective when it explicitly targets specific types of benefit opportunities that will be provided at designated locations. This is done by providing activity and associated setting opportunities defined in terms of the beneficial experiences and other responses that can be realized from using those opportunities. For example, site A might be targeted and managed in part to provide opportunities for physical fitness, with chin-up bars, climbing equipment, etc. Site B might be managed in part for learning about a cultural/historic site or for promoting a better understanding of natural ecosystems; site C might be managed in part for enhancement of self-concept and identity through the development and application of specific skills; site D for family cohesion, and so on.

BBM focuses on what is obtained from amenity resource opportunities in terms of consequences that maintain or improve the lives of individuals and groups of individuals, and then designs and provides opportunities to facilitate the realization of those benefits. The basic purpose is to provide an array of benefit opportunities among which users can choose. Several benefit opportunities can be targeted for the same site or area.

BBM can also be experience-based management, simply because many of the benefits of leisure (e.g., a positive change in mood, enhanced self-concept, increased family or group kinship) are affective responses, generally of short time duration. EBM is usually BBM because the experiences people seek are desirable and beneficial but BBM specifies formulation of management objectives that target outcomes that are positive and beneficial.

Figure 3 illustrates the role of resource managers in producing recreation experience and benefit opportunities and how recreationists use those opportunities to produce benefits for themselves and society. We emphasize that managers

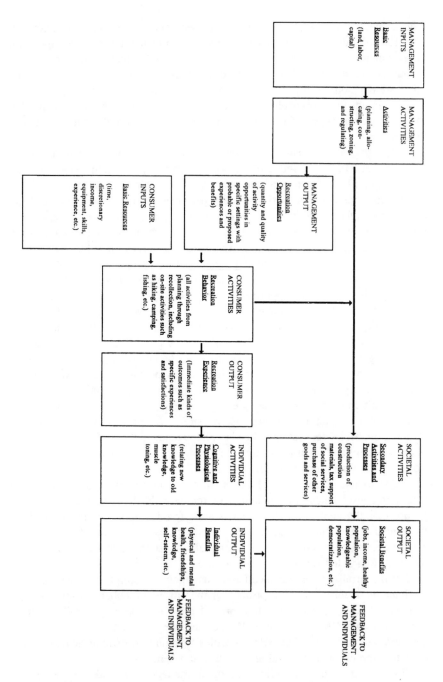

Figure 3 Overall process and subprocesses for producing outdoor recreation benefit. (Reprinted with permission of the author (Brown, 1984).)

150

do not provide recreation experiences or benefits; their products instead are opportunities. Users produce experiences and benefits for themselves by using those opportunities and integrating their responses into their lives both during and after (sometimes a rather long time after) participation.

BBM provides managers with a clearer understanding of the outputs of the amenity resource management process, enabling them to identify specific management objectives that explicitly specify the types of benefit opportunities to be provided, where they will be provided, when, and in what amount.

Up to this point, we have largely focuses on only one of the products of BBM: those benefits accruing to users or those who take advantage of benefit opportunities. Figure 3 also illustrates a second by-product of BBM: provision benefits. Provision benefits are social and economic benefits that result from public investment in the provision of amenity goods and services. For example, agencies who provide opportunities for people to engage in recreation on public lands also create jobs, income, and tax support for local communities and regions.

CURRENT STATUS OF BENEFITS-BASED MANAGEMENT

Support for BBM continues to grow. The Rocky Mountain Station is pursuing research on the benefits of amenity goods and services on two supplementary and complementary fronts, one basic and the other applied.

The direction of the basic research focuses on (1) clearly specifying the types of benefits that are likely to or are known to accompany the provision and use of clearly defined amenity goods and services, and (2) measuring the magnitude of these benefits for well-specified market segments of beneficiaries (either communities or individuals). The basic research is organized into subareas of inquiry that concentrate either on a particular class or type of benefit (e.g., spiritual benefits, benefits associated with different air ion concentrations, learning benefits, or promoting an environmental ethic) and determining the role of natural environments in promoting benefits, or on developing methods and instruments to measure and quantify other benefits of amenity goods and services.

CASE STUDIES

The Rocky Mountain Station has supported applied research that focuses directly on refining and applying the concept of BBM that emerged from the Estes Park applications workshop. This research centers on agency and university scientists helping to design and evaluate pilot test applications of the BBM concept to different types of amenity resources. At least two pilot projects are currently under way, one with the city of Portland, Oregon, and the other with the Bureau of Land Management (BLM) in Grand Junction, Colorado. Rocky Mountain Station scientists and potential collaborators in public recreation agencies are also considering other pilot tests of BBM. Those efforts are contingent on funding, and sponsoring agencies include Minnesota State Parks and Recreation, the

Western Region of the Canadian Park Service, the Roosevelt and Arapaho National Forests in Colorado, and the Coconino National Forest in Arizona.

The approach taken in the pilot studies is for researchers to work closely with managers using both judgment and research to identify the types of benefits to individuals or groups that can be associated with a particular resource area. Once those benefits are identified, management objectives are written that explicitly target provision of opportunities for resource area users to achieve those benefits. The BBM pilot project currently under way in Grand Junction, Colorado, illustrates this process.

The Grand Junction pilot is being carried out on the Ruby Canyon–Black Ridge (RC–BR) area in eastern Colorado. Managed by BLM, the area contains cultural, paleontological, river recreation, mountain biking, dispersed recreation, and wilderness resources that could provide a variety of benefits associated with learning, cultural appreciation, and spiritual values realized by both individuals and groups. The area may also provide economic and quality of life benefits associated with recreation opportunities for local and nonlocal users.

In designing the RC–BR pilot project, researchers and managers sought to gather information about the area's individual, group, and community benefits from a number of sources. The first step was to divide the RC–BR area into management zones based in part on the ROS inventory of the area. Managers then participated in an exercise where they compiled a list of individual and group benefits that they perceived were being provided to those who recreated in each of the management zones and the community benefits that were being realized by nearby Grand Junction.

Two strategies were used to assess the benefits as perceived by RC–BR users. The first was a survey of on-site visitors designed and conducted by managers and cooperating researchers. The study utilized both open- and closed-ended questions to assess user-defined benefits and the activity and setting attributes that contributed to realization of benefits within the RC–BR manager-defined zones. In addition to the on-site visitor survey, focus group interviews were conducted with local area recreation user groups such as the mountain biking club, jeep club, equestrian group, rock and mineral club, and the local chapter of the Audubon Society. The open-ended responses to questions about individual and group benefits provided by these user groups were compiled and provided a second source of information on user-defined benefits.

A sample of Grand Junction community leaders was surveyed to provide a third source of information on benefits. This group was asked to identify the economic and other community benefits associated with recreation use of the RC–BR area by local residents and visiting tourists.

With benefit information gathered from on-site users, local user groups, and community leaders with their own judgments on RC–BR area benefits, managers moved to the next stage of the pilot project and began the complex task of translating the wealth of benefit data into BBM objectives for the RC–BR area. These objectives will define specific types of benefit opportunities to be targeted in each of the area's management zones. This phase of the pilot is still under way. Managers are also talking with Grand Junction community leaders and recreation "partners" who provide goods and services to area recreationists about the community's role in providing benefits to visitors and the community.

The Grand Junction pilot project is an ambitious and challenging undertaking by committed BLM managers and cooperating researchers to implement BBM from the benefit-identification stage to writing and managing for targeted benefit opportunities. When it is completed, managers and researchers will document the entire pilot process used at RC–BR to serve as a test and model for other public land managers seeking to implement BBM.

A BBM Interagency Steering Committee was formed in 1993 at the request of the chief of the National Resource Management Division of the Army Corps of Engineers and the heads of the recreation staffs of the U.S. Forest Service and BLM. The committee is made up of representatives from federal land management agencies, directors of state and municipal recreation agencies, university researchers, and representatives from the National Recreation and Park Association, the American Recreation Coalition, and private recreation and leisure providers. The charge is to develop and guide an implementation plan for BBM.

BBM is in the evolutionary and developmental stages; it will continue to progress as a process of researchers learning and exploring with managers. The applied focus in the development and management application of BBM offers great potential for managers trying to cope with today's changing resource management issues. The ultimate beneficiaries of BBM will be all of society as agencies learn to identify and manage for those opportunities that deliver the greatest benefits.

ENDNOTE

1. This paper was written in 1992. Since that time, the thinking about benefits-based management has progressed. The following changes are particularly relevant:

The definition of recreation benefits has been expanded to changes that are viewed to be advantageous or improvements in condition (gains) to individuals (psychological and physiological), to groups, to society, or even to another entity; and the realization of desired and satisfying on-site psychological experiences.

Benefits-based management is also being taught at the Bureau of Land Management's National Training Center.

Additional pilot projects are under way in Chandler Park in Detroit, Mount Rogers National Recreation Area in Virginia, in South Carolina, and on the Los Caminos Antiguos Scenic and Historic Byway and the Alpine Loop Back Country Byway in Colorado.

REFERENCES

Allen, L. R. 1990. Benefits of leisure attributes to community satisfaction. *Journal of Leisure Research* 22(2): 183–196.

Brown, P. J. 1984. Benefits of outdoor recreation and some ideas for valuing recreation opportunities. In *Valuation of wildland resource benefits,* edited by G. L. Peterson and A. Randall, 209–220. Boulder, Colorado: Westview Press.

Driver, B. L., and P. J. Brown. 1987. Probable personal benefits of outdoor recreation. In *A literature review—President's Commission on Americans Outdoors,* 63–70. Washington, D.C.: Government Printing Office.

Driver, B. L., T. C. Brown, and W. R. Burch. 1986. A proposal for more comprehensive and integrated evaluations of public amenity goods and services. Presented at the Congress of the International Union of Forest Research Organizations, Yugoslavia.

Driver, B. L., P. J. Brown, G. H. Stankey, and T. G. Gregoire. 1987a. The ROS planning system: Evolution, basic concepts, and research needed. *Leisure Sciences* 9(3): 201–212.

Driver, B. L., R. Nash, and G. Haas. 1987b. Wilderness benefits: A state-of-knowledge review. In *Proceedings—National wilderness research conference: Issues, state-of-knowledge, future directions,* R. C. Lucas, compiler, 294–319. Gen.Tech, Rept. INT-220. Ogden, Utah: Intermountain Forest and Range Experiment Station.

Driver, B. L., and G. L. Peterson. 1987. Benefits of outdoor recreation: An integrating overview. In *A literature review—President's Commission on Americans Outdoors* 1–10. Washington, D.C.: Government Printing Office.

Ellis, T., and G. Richardson. 1991. Organizational wellness. In *Benefits of leisure,* edited by B. L. Driver, G. H. Peterson, and P. Brown, 303–329. State College, Pennsylvania: Venture Publishing.

McDonald, B. L., and R. Schreyer. 1991. Spiritual benefits of leisure participation and leisure settings. In *Benefits of leisure,* edited by B. L. Driver, G. H. Peterson, and P. Brown, 179–194. State College, Pennsylvania: Venture Publishing.

Orthner, D. K., and J. A. Mancini. 1991. Benefits of leisure family bonding. In *Benefits of leisure,* edited by B. L. Driver, G. H. Peterson and P. Brown, 189–301. State College, Pennsylvania: Venture Publishing.

Roggenbuck, J. W., R. J. Loomis, and J. Dagostino. 1990. The learning benefits of leisure. *Journal of Leisure Research* 22(2): 112–124.

Rolston, H., III. 1985. Valuing wildlands. *Environmental Ethics* 7:23–48.

Ulrich, R. W., U. Dimberg, and B. L. Driver. 1990. Psychophysiological indicators of leisure consequences. *Journal of Leisure Research* 22(2): 154–166.

Chapter 11

Parks Canada's Economic and Business Models: Perspectives on Their Development and Use

Jay Beaman

Luc Perron

Dick Stanley

For over thirty years, Parks Canada has used economic models to make decisions about new parks and new capital development, as well as for other planning purposes. These models are based on economic theory that is well developed in the literature. The references in the documents cited at the end of this chapter provide the reader with a basis for examining that theory as it has been applied to the model in question. This chapter concentrates on those organizational reasons that led Parks Canada to develop and automate the models.

Parks Canada spends about $400 million (Canadian) every year. These expenditures are associated with over twelve thousand built assets and hundreds of kilometers of paved highways. It provides 1.6 million site nights of camping annually and receives twenty-five million person visits. Though there are various estimates of the worth of Parks Canada's assets, a useful reference figure is $5 billion. During peak season, up to ten thousand employees are involved in delivering service. Services are delivered at 34 national parks, 70 historic sites, and 9 historic canals. Because of the size of its operations, Parks Canada recognized in the 1960s the need to use economic models and automation to help it plan and manage its budgets, assets, and services.

Figure 1 provides an overview of the history of Parks Canada's economic models. The figure and the somewhat longer list of references at the end of this chapter are only partial. References are omitted either because they are merely minor variations on cited models and present no significant interest, or because they involve internal government business that is not public. The references cited at the end of this chapter nevertheless give a full picture of history of Parks Canada's model development. Some of the automated models themselves are available to the public and are, in fact, in use at universities (e.g., University of Waterloo and Texas A&M). However, much of the software is dated, and some of it has proprietary restrictions over its use. High-level languages now available produce nonproprietary output code so that models could now be available at a very low price, but when these models were developed, we chose to save on development by using proprietary products. The cost of buying software that permitted us to distribute the models was far cheaper than writing our own code. The authors are therefore not suggesting that readers should obtain particular Parks Canada models. Still, many readers may benefit from obtaining user and systems manuals for models or papers on the results of using these models. Because of governmental reorganizations, the organizational parts of many of the references are out of date, so one should not try to obtain Parks Canada documents based on information in the list of references.

WHY MODELS AND WHY AUTOMATE?

It is easy today to think of a model as something that one runs in a statistical package or that one quickly builds using a spreadsheet. The models treated in this chapter are much different. They are not built ad hoc in a statistical package and then erased. These models are well documented and are of a more permanent and user-oriented nature.

They are rigorous enough that they can be used as defacto standard methods or approaches to problems. If a large number of people carry out a large number of separate economic or business analyses, the result can be a lot of analysis done in different and often unknown ways. A model as an analysis standard provides a large organization with the kind of uniformity that is necessary if it is to hope to have comparable analyses. There are a number of variations an analyst can introduce, such as the use of different categorizations of capital expenses, different input-output tables or alternates for these, or different computational procedures. The purpose of a de facto standard is to define the ground rules that must be met or exceeded for a given analysis to be organizationally or scientifically acceptable. With a standard defined by a paper procedure or by an automated model, it is possible for anyone to know how results were arrived at, and it becomes easy to document deviations from the standard. This leads to logically consistent and reproducible results for park planning.

Automated models also promote accuracy. One aspect of accuracy is getting the "right" number and carrying out the appropriate computations correctly.

Year	MAJOR ECONOMIC STUDIES AND MODELS DEVELOPMENT
1964-70	* Economic Survey of the Kejimkujik Park Area in Nova Scotia. * Economic Impact Study of Alternative National Park Proposals at Val Marie, Saskatchewan and Riding Mountain National Park. * The Economic Impact of National Parks in Canada, A Theoretical Framework for the Evaluation of the Benefits and Costs of National Parks (+ the Gros Morne NP Case Study).
1974-81	* Regional Socio-Economic Impact of a NP: Before and After Kejiimkujik. * Evaluation of the Impact of a National Park: The Longitudinal Study * Analysis of Socio-Economic Impacts of the Proposed Grasslands NP and Regional Economic Analysis of Cape Breton Highlands NP. * Socio-Economic Impact of Kouchibouguac NP on its immediate environment and on the Province of New Brunswick prepared for Special Commission of Inquiry (Federal-Provincial).
1983	NATIONAL ECONOMIC IMPACT METHODOLOGY * Evaluation of the Impact on the Canadian Economy of Spending Attributable to National Parks and Sites Administered by Parks Canada.
1984-86	ECONOMIC IMPACT MODELS AND THEIR AUTOMATION (Softwares used: BASIC and wylbur unless otherwise specified) * VISIT EXPENDITURES MODEL (not fully automated; SAS) * Beaman J., The Parks Canada Expenditure Model. * CAPITAL PROJECT ECONOMIC IMPACT MODEL (automated in 1984) * LOCAL (TIEBOUT) ECONOMIC IMPACT MODEL (automated in 1985) * LIFE CYCLE COST (DISCOUNTING) MODEL (automated in 1986) * PROVINCIAL ECONOMIC IMPACT MODEL (automated in 1987)
1987	EXTERNAL REVIEW OF ECONOMIC/BUSINESS MODELS * P.G. Whiting and Associates, User Requirements Study and Systems Manual for the Economic Models of Parks Canada.
1986-87	* Socio-Economic Analysis for Project Planning: Participants Manual.
1987-92	TERRITORIAL AND PROVINCIAL ECONOMIC IMPACT STATEMENTS (based upon the methodology developed in 1987 for the PEI model) * Yukon and NorthWest Territories * Alberta * Atlantic Region (Provinces of Newfoundland, Nova Scotia, Prince Edward Island and New Brunswick) * Ontario * Prairies (Provinces of Manitoba and Saskatchewan)
1989-94	DEVELOPMENT OF ADDITIONAL ECONOMIC/BUSINESS MODELS * MAKE OR BUY MODEL (automated in 1989; Lotus 123) * INTEGRATED BUSINESS INFORMATION SYSTEM (automated in 1993-94; Access) * REVENUE ASSISTANT MODEL (automated in 1993-94; Toolbook)
1995-98	* ECONOMIC VALUATION FRAMEWORK OF PROTECTED AREAS * REDESIGNING PROVINCIAL ECONOMIC IMPACT MODELS

Figure 1 History of Parks Canada's models.

Left on their own, many people who need results and could do the computations do not do them enough to be able to quickly and correctly select data from published sources. They may also make errors in selecting computational formulae. Many modeling initiatives involve a significant number of calculations. When such modeling began in the 1970s and even well into the 1980s, all work was being done with manual calculators and using paper. A typical capital project analysis, when done manually, involved a day to a day and a half of work. Also, it was found that most modeling exercises had to be repeated in a systemic way just to find and eliminate serious clerical and computational errors. With an automated model, computational errors are eliminated, and the model does the computations in seconds, not days.

If an organization places importance on knowing the economic impacts of new and existing parks, the economic impacts of hundreds of major capital proposals, or the life cycle costs of many large projects, a large amount of time and financial resources is going to be needed. Person power is required to produce accurate results with a known and consistent methodology. When even a small project is analyzed manually, more than a person-day of work is involved for someone who knows what they are doing. Automated, the task takes an hour or so for somebody who is familiar with the software, and this does not mean employing an economist or a person trained to do the computations "from scratch." Automation quickly pays for itself by saving in staff salary or by avoiding fees for somebody else to do the work.

A final consideration in support of automated models is their usefulness in training. The model can be a learning tool and can even be used for computer-assisted learning (CAL). Certainly, the Parks Canada models have all been viewed as important learning tools, though only the revenue estimation assistant model was developed with this as a specific goal. Other models have been provided with user guides to teach the user how to use the model and interpret the results obtained.

PARKS CANADA MODELS AND CHANGES IN COMPUTING

Parks Canada could not automate its models until appropriate electronic data processing (EDP) tools were available. Using the kinds of models described here by means of punch cards or "typewriter" style terminals to input to a mainframe computer did not work. Low-speed communication and very limited access to terminals, key punchers, and slow printers were enough to discourage automation. They were enough of an impediment to discourage use of computers for many purposes. Even as personal computers (PCs) appeared in Parks Canada, access was so limited that applications could not be put on them until the mid-1980s. The initiatives early in the 1980s to automate a national economic impact model did serve to focus on problems of translating a paper procedure into a set of consistent equations. In addition, it showed that existing software and hardware presented problems of data storage and storage of results, as well as development of convenient user interfaces.

PCs changed the situation radically and quickly. Parks Canada began their microcomputer modeling work with a 64K PC with two floppy drives and a ten megabyte hard disk (an XT). What one could do was severely limited by both the hardware and software available. In the models developed in the 1980s, one sees these limitations. These are no Windows applications or even applications for color monitors.

In the 1980s, Parks Canada found some solutions using powerful database software developed for XTs, but it was the "new" computer technology, the 286, that really facilitated developments in modeling. Even with the 286, many problems such as availability and reliability of software, "speed," and effective use of memory diverted research resources from the substance of problems to "Can we?" or "How do we get around?" With 386s, large disk storage, networks, and powerful software, incredible things became possible. What the analyst does and how he does it is no longer limited by technology as it was in the 1970s and early 1980s.

PARTICULAR MODELS AND THEIR EVALUATION

Economic Impact Models

National Economic Impact Model   Parks Canada's first National Economic Impact model was developed very quickly in the late 1970s. It was necessary to provide evidence of economic impact to defend Parks Canada's budget against cuts. The first national model and impact statement was created by "locking up" two people (one of the authors included) for two weeks to do computations while others provided data and moral support. To avoid a second lock-up, Alain Charlebois from Parks Canada's Quebec region came to Ottawa for an extended period to create a systematic model and produce a more rigorous economic impact statement (Parks Canada, 1983). Manual computations of operating costs and visit expenditures were done in parks, and the expenditures were apportioned using Tourism Canada's Tourism Expenditures Model (Bureau of Management Consulting, 1975). Statistics Canada's Input-Output model of the Canadian economy was used to produce the final estimates. As a consequence, the Parks Canada results have credibility based on the credibility of these other organizations.

Although the Charlebois approach produced reasonable results, the real problem was getting results regularly updated. The approach was sufficiently laborious and consumptive of resources to preclude applying it manually year after year. In 1984–1985, therefore, we began to computerize this work on a mainframe. Having computing power available quickly showed that more than two person-years of computational work could be accurately done in minutes. Of course, getting, organizing, and inputing data took several weeks, and then getting work done by Statistics Canada took further time. Nevertheless, computerization made it feasible to do the updates regularly.

Computerization brought computational speed, and it also brought the opportunity to formalize the analysis rigorously. Doing the computations manually in a reasonable time had meant compromises in certain formulae, for example, in how length of stay and proportions of users by origin were treated. With a computer, some of the implications of formulae could be studied by creating data sets of implied behavior and analyzing them. Such examination showed that some of the manual approximations to reality were not very good. As a result, automation efforts turned to a new model, the Visit Expenditure Model described below.

Since 1985, the National Economic Impact Model has been updated but not by a standard procedure. Results of the 1987–1992 studies (cited in Figure 1) have demonstrated that visitor expenditure data derived in the regional/provincial impact statements are more reliable than estimates developed from parks attendance and camping statements based on formulae such as those used in the 1970s and early 1980s. Therefore, in the interest of resource savings, there is no national economic impact model as such. The update of the national impact now depends on the completion of regional/provincial level studies and appropriate corrections for national accounts.

Capital Project Economic Impact Model   Parks Canada spends about $100 million a year on capital projects. Given the concern with expenditures from the late 1970s to the present day, significant resources were devoted to computing economic impacts associated with capital expenditure on a project-by-project basis.

Considering the cost of doing such work, the variability of approaches, and the potential for errors, it was quickly recognized that a standard and less costly approach was needed. As a result, in 1984 the Capital Project Economic Impact (CPEI) Model was the first model automated by Parks Canada on a PC. It is based upon Statistics Canada interprovincial input-output model (using 1984 data). It has been used extensively by Parks Canada for a decade to produce reliable estimates of the provincial and national economic contributions resulting from major capital investments. With several hundred capital projects per year for which impact statements were to be produced, automation allowed the skilled person to use the power needed to do computations manually for other work. Analyses were being done in five regions and some parks. A standard was required to get accurate results produced in a consistent manner.

The model quickly came to play an important role in educating staff about economic impact without them getting involved in finding input-output (I-O) tables, learning about industry classification, making extensive computations using worksheets, or learning which formulae to apply.

Local Economic Impact Model   The economic impact of parks has been an active area of analysis for thirty years.The Local Economic Impact (LEI) Model (*User Guide,* June 1985) is based on Tiebout (1962). It allows an analyst to define the economic contribution of a national park, site, or canal to local and regional economies by providing specific data to the model. The model then provides an estimate of the structure of the area studied (interindustry linkages and leakages). This model has been the de facto standard for all local/regional economic impact studies by Parks Canada since its development (see Figure 1). This model in the mid 1990s remains in use by regional and headquarters analysts (as well as by consultants and universities in the United States, Canada, and Australia). In Parks Canada, it is used for planning for existing parks and to study the anticipated economic impacts that might be associated with the establishment of new parks and site proposals.

The LEI was introduced for a number of reasons. It is a productivity tool. The model saves a lot of resources over manual calculation. It does not require much quality assurance time by a knowledgeable economist. These are important savings, but possibly more important was having the LEI as a standard. There are any number of park impact studies that give noncredible estimates of benefit. In the 1970s and 1980s, studies used "multipliers" of 2 and more for all kinds of purposes. Often it was not even clear what the impacts really meant in rigorous economic terms even when they were reported in dollars and person-years of employment. Parks Canada requires the LEI be used or that equivalent results be produced in a documented manner (using the LEI as a standard). This has resulted in consistent and comparable economic impact results.

Provincial Economic Impact Model   The province is the political unit that the Canadian government must deal with on a variety of matters having to do with parks. For example, new park creation typically involves provincial participation in land acquisition. The Canadian government can often benefit from explaining park's economic contribution to the province. This is where the Provincial Economic Impact (PEI) Model comes in.

One can view this model as an LEI model with a province as the local area. However, because there are provincial I-O tables, the model is based upon coefficients obtained from Statistics Canada. When these provincial I-O coefficients are combined with good estimates of expenditures by visitors from within and outside the province, the model can be used to compute the economic impact of a particular park on the province that it is in.

This model has been used extensively by Parks Canada in the preparation of a series of major studies entitled "Visit Profile" and "Economic Impact Statements" prepared for all provinces except British Columbia and Quebec. These studies provide baseline socioeconomic information that is useful for socioeconomic analyses conducted during the different stages of the management and capital project planning processes. They help to raise support for parks in general and indirectly create a favorable attitude toward the establishment of new ones. These studies provide senior managers and analysts with facts on socioeconomic matters associated with the program to help prepare, for example, notes for formal public meetings and information for new national park and historic site negotiations as well as for use in preparing marketing and business plans.

Visit Expenditure Model and Park Use Related Data System   The Visit Expenditure Model (VEM) has only been automated as a prototype and so has not advanced beyond a methodology or framework. As indicated earlier, VEM arose in part out of limitations in the Charlebois approach to national economic impact and in part from the question, If you have general expenditure data on domestic and international travelers, how do you use these to get better results in specific studies? The answer is that you cannot. The model was built to get estimates with no such data available. Rather than providing assumptions about desegregation on such characteristics, the model uses assumptions that do not allow unique decomposition, nor for detailed data to be used to improve results from the model.

By 1985–1986, Parks Canada had gained a great deal of experience with providing the kind of information that planners and managers needed. Parks Canada also was conducting a large number of visitor surveys and getting economic data on park visits. However, there was no systematic framework for analysis of data in general or visitor economic impact and visit expenditure in particular. The Park Use Related Data System (PURDS) was introduced as a general framework for data, and the VEM was introduced as the framework for visit-related economic data (Beaman and Grimm, 1989; Stanley et al., 1992). The goal in VEM was to provide a sound basis for recognizing three things: expenditure-sensitive user segments, various accounts that expenditures might be associated with, and various economic "sectors" or expenditure types that would be compatible with Tourism Canada and Statistics Canada data collections and analyses. The first version of the VEM was called the Visitor Expenditure Model. It was subsequently recognized that visitors may, as many do, make different types of visits, so the model was reworked as the Visit Expenditure Model. A variety of issues dealing with individual visits and expenditures by persons or parties had to be addressed in the new VEM.

Both VEM and PURDS are discussed below based on an overview by Beaman et al. (1994). The most recent applications of the principles of the VEM as a methodology resulted in our best estimate of the amount spent by visitors in Canada, attributable to their visits. In 1994, national heritage locations managed by Parks Canada received twenty-five million person-visits that resulted in expenditures of nearly $1.4 billion in the surrounding regions where the parks and sites are located. They represent nearly 4 percent of the total tourism spending in Canada. Park and visitor related expenditures ($1.8 billion) help diversify the economies of remote regions and contribute to the stability of incomes and employment. The 1994 estimates of the contribution of Parks Canada to the Canadian economy is evaluated at over $2 billion worth of GDP and helps to sustain nearly fifty thousand full-time jobs annually.

Business-Oriented Economic Models

The models reviewed up to now have centered on economic impact. With government budget restraints in the late 1970s came an acceptance that decision support systems that facilitated more effective operations were necessary. The following provides information on several Parks Canada's initiatives in this area. The effectiveness of models is also addressed in the discussion section.

Project Options Analysis (not automated) This analytical approach (Parks Canada, 1987) is an elaboration of CORDS Technical Note No. 25 (Acar, 1976). Sponsors of projects in Parks Canada repeatedly face the problem of assessing options in terms of their relative ability to meet predefined objectives. The problem that managers face is rating and ranking various options using both quantitative and qualitative factors.

Project Options Analysis represents a refinement of cost-effectiveness analyses. The main focus is on intangible factors. Many options for development involve visitor satisfaction, regional integration, and aesthetics appreciation. Often, when scales were developed and scores assigned, the same factor would end up being counted in many different variables, thus distorting the score. The Project Options Analysis specifies an approach whereby intangibles can be integrated into a cost-benefit decision-making process. This analysis was not, however, a good candidate for automation in the 1980s because the scaling could be quickly done using paper or a blackboard.

Project Options Analysis has merit both conceptually and practically. If managers or other staff are going to rate intangibles, then the principle enunciated in the Project Options Analysis should be observed. However, in the 1990s, many problems viewed as qualitative in the 1970s and 1980s can be approached quantitatively and automated. Choice modeling, conjoint analysis, contingent valuation, etc., offer alternatives to ad hoc scaling based on arbitrary judgements.

Life-Cycle Cost Discounting Model As project analysis became more rigorous and budgets more restrained, a business-like approach to present and future costs became part of the analysis. This model was developed to provide a standard discounting analysis for accumulating data and computing present values of future cost and revenue streams. It has also served as a learning tool whereby staff could, in a two- or three-day course, become familiar with life-cycle costing.

The tool is a discounting model that allows users to compare the present values of the costs and revenues of the various design and delivery options and to evaluate investment scenarios over a common period of analysis (a life cycle). It thus helps managers to make decisions through identifying the lowest cost solutions, determining whether a current investment justifies a future cost saving, and finding out when it will be more economical to replace an asset than to repair it.

Revenue Estimation Assistant Model With the need in the 1990s to increase revenues, it has become critical to analyze the effects of new fees or other types of charges on client groups. The revenue assistant model is a tool that provides a structured approach for making revenue estimates for a park's services. Changing park entrance fees, camping permit fees, and other fees for private benefits does not automatically mean more revenue. Higher fees may, in fact, result in lower use levels and lower revenue.

The model, which is based on sound market research and economic principles, guides the user to consider the factors that determine whether a fee hike will increase or decrease his revenue and by how much. It asks the user to define various price-sensitive visitor market segments (and their relative sizes). Based on the segment sizes, coefficients of price sensitivity set for them, and trends in use by them, the model computes impacts on revenue. Through a dialogue process, the model takes the user through each of these basic issues and prompts him to incorporate whatever data he has, or make informed judgments in the absence of hard data. It thereby also serves as a learning tool.

Revenue estimation is one part of a range of considerations that need to be taken into account in an integrated fashion. Raising fees may impact the relative use of a park by different segments of the population. This has consequences in terms of the socioeconomic impact a park has on its region. Fees also impact the use of different facilities "independently" of changes in user groups. User groups may change their use patterns. To deploy person power effectively for operations and to spend wisely on recapitalization and new capital, it is necessary to plan considering the fees to be charged and their consequences. So, this model is part of a larger costing and business planning picture discussed below.

Make or Buy Model   The Make or Buy Model, like the Life-Cycle Cost Model, was introduced to establish a standard. It provides an analytical framework for the evaluation of make or buy options according to financial and accounting principles. It is used in planning for service delivery. It estimates comparable costs for Parks Canada to provide a service in-house or through outside means. Since its automation in 1990, it has been used mainly as a training tool for staff involved in contracting such work and as a reference for specifying to contractors how Parks Canada expects make or buy analyses to be done.

The Integrated Business Information System   With the advent of business planning in the 1990s, it has become evident to Parks Canada managers that they need information on costs of services and the impacts of those services. Initial attempts to use existing systems to get this information have shown that the necessary systems either do not exist or contain data that have to be transformed manually before being used. Managers therefore looked for some automated means to produce the cost and consumption reports they needed to manage their operations as a business.

The Integrated Business Information System was developed in response to this need. It is a database application that electronically combines data from the Parks Canada financial accounting system, various use information and inventory (miles of road, number of campsites) databases, etc., to produce reports for managers on cost per unit of production or per unit of consumption. The model sets up predefined reports so that the manager can obtain cost and performance information about his or her operations without having to do laborious studies like those done when business planning was first established.

This model reports on past expenditure (which is captured in the financial system) and past performance (as recorded in whatever other systems exist). It does not project costs or performance into the future, although records of the past are obviously indispensable for any projection.

DISCUSSION

Some of the models described here, such as the VEM, were never developed beyond the conceptual or prototype stage. Some of the models that were fully developed, such as the CPEI, and that were used for innumerable projects, are now dated, not in terms of economic theory, but in terms of EDP equipment for which they were written and the capability of the software products used. It is to be expected that with the expanded computer and software capabilities of the 1990s, some of these models will be revisited and rewritten.[1]

One of the main trends of the 1990s is the increasing ability to support large integrated data sharing systems that are powerful enough to provide artificial intelligence, expert systems, and decision support (Hamilton-Smith and Mercer, 1991). Furthermore, Parks Canada is moving into a period where rational business decision making is becoming important as never before, and such decision making requires the support of information and analysis of the kind only today's computer systems can deliver.

The current state of the models, whether frozen in the prototype stage or built on dated platforms, presents Parks Canada with a unique opportunity. The prototype models such as VEM and PURDS foresaw the current computer environment and contain the principles necessary for successful large-scale integration and use of data in ways that are possible now but were not when the models were developed. The dated models such as the LEI and the PEI still contain the kind of analysis that is needed for decision support; only now can their integration into overall decision support tools or expert systems be contemplated. It is now appropriate to bring tourism, visitor expenditure, facility service operations, and capital cost information together into one affordable database. Routine updates of a variety of statistics are possible, so that new national economic impacts are part of a reporting procedure. However, this step requires a variety of information integration and definition initiatives that are now proceeding within the Department of Canadian Heritage. An effective, easy to use decision support tool based on these models is within our grasp.

All models were developed in the context and full awareness of the way the discipline was evolving. For example, a major external review and evaluation of our models was done in 1986 by Whiting. In 1989, the Conference Board of Canada invited Parks Canada researchers to present their models at a national conference. Even currently emerging themes like continent valuation of natural resource benefits have found an echo in our model development. In the 1970s and early 1980s, models did not involve such methods for valuing nonmarket benefits of parks and recreation because managers did not find the arguments compelling. However, with the renewed interest in contingent valuation in the discipline and the success of arguments based on this method in the courts in the United States, researchers in Parks Canada have been examining ways to introduce the consideration of economic benefits into what has been mainly a debate around economic impact.

CONCLUSION

All three authors can look back at each of their two decades of involvement in the development of economic and business models in Parks Canada. These developments certainly generated a kind of excitement. There was also the drudgery and effort that went into such activities as acquiring technology, getting the systems working, educating users, and obtaining and justifying funding.

Looking back, it is possible to judge the success of the modeling effort on the basis of four criteria: standards, accuracy, effectiveness, and learning tools. Under these criteria, the models have served Parks Canada well. There are standard arguments against developing models, such as "every problem is unique," "I need it yesterday so there is no time," "I can do this study cheaper without a general model." The authors believe that Parks Canada's experience has shown that devoting a reasonable level of resources to proactive development of models has been a good investment. A key consideration is that the resources spent should offer a reasonable payback in terms of savings from using the model. In such work, it is important to recognize that a framework or manual analysis procedure that never becomes an automated model will still have a payback as a de facto standard and for use in training.

It is also possible to look back and see that having a de facto standard has resulted in quality work with comparability of results within the organization over time. As for efficiencies and savings, one could estimate the years of staff time and millions of dollars in consultant fees that have been saved by carrying out computation using models rather than manual approaches. The models cost a small fraction of what was saved by using them. As for models and learning, Parks Canada has developed a wide range of staffs who have a good basic awareness of what goes into models and what comes out. These staffs have examples of why and how models are to be used. Having more and more knowledgeable staff users of models has contributed to the development of organizational awareness of economic impact, options analysis, and other areas important to improving Parks Canada's performance as a business.

Certainly, if there is a related area in which it would have been desirable to see more progress, it is the integrated use of information. The Integrated Business Information System and the Park Use Related Data System (including the Visit

Expenditure Model) are examples. Economic impact models and business models covered here are also part of the integrated picture. The key to progress in integration is that data must be treated in ways that allow for generalization, transferability, and comparison. It is not enough to have models that let computations be done in a standard way and produce results that are comparable from one location to another. As one probes more deeply into the management uses of information and sees what needs to be considered in terms of different types of accommodation, where travelers come from, etc., one sees the need for a data structure that is both complicated and systematic. This structure is the bridge that integrates models and allows a really effective Decision Support System. It is the authors' hope that by the year 2000, significant progress has been made toward an integrated Parks Canada decision support system.

ENDNOTE

1. Since the writing of this paper, the PEIM has, in effect, been redesigned to reflect emerging trends in hardware and software products (Windows environment, Pentiums, Visual Basic 4.0, and MS Access). Previous economic impact models were no longer operational (developed in a DOS environment). Previous capabilities of the old Capital Project Economic Impact (CPEI) and Provincial Economic Impact (PEI) models were fully integrated and expanded into this new version of the model. It also reflects the broader concerns of the Department of Canadian Heritage, Parks Canada's new parent department.

The model offers analysts a flexible tool with which to evaluate the provincial economic impacts of museums and historic sites, cultural events and festivals, major games and other sporting events, as well as protected areas such as national and provincial parks and heritage rivers. The model uses up-to-date coefficients and multipliers from Statistics Canada Interprovincial Input-Output Model that make it possible to trace the effects of an increase in x dollars in the demand for a specified commodity through the provincial economies, in terms of GDP (value added), labor income, and level of employment (number of full-time equivalents). The model includes a standardized set of expenditure categories (each has its own set of impact coefficients) for each source of economic impacts considered. The model takes into account (1) operating and maintenance expenditures (including the payment of wages and salaries and fees paid to artists), (2) infrastructure expenditures (including expansion of existing facilities, such as adding a new wing to the Art Gallery of Ontario or expanding a visitor reception center in Banff National Park), and (3) attributable provincial spending made by tourists (visitors) while visiting a park, historic sites/museums, or cultural festivals.

CHRONOLOGY OF STUDIES

Major Economic Studies and Model Development: Chronology of Studies

1964   Schafer, Paul D., and Robert D. Comeau, Economic Survey of the Kejimkujik Park Area in Nova Scotia, The Institute of Public Affairs, Dalhousie University.

1965   Morse, Norman H., An Economic Evaluation of a National Park, Acadia University Institute.

1968   Acres Research and Planning Limited, An Economic Impact Study of the Proposed Bloodvein National Park.

Kaplan Consulting Associates Ltd., Economic Impact Study of Alternative National Park Proposals at Val Marie Saskatchewan.

Wise, Gladstone and Associates, Economic Impact of Riding Mountain National Park.

1970   Hildebrandt, Young Associates Ltd., The Economic Impact of National Parks in Canada, Volume 1: A Theoretical Framework for the Evaluation of the Benefits and Costs of National Parks. Volume II: The Evaluation of Benefits and Costs of Proposed National Parks, the Gros Morne Case.

1973   Parks Canada, Economic Impact of the Proposed Ship Harbour National Park on the Eastern Shore of Nova Scotia.

1974   Beaman, J., L. Lehtiniemi, and R. Stanley, Evaluation of the Impact of a National Park: The Longitudinal Study.

1975   Mayes, R. G., Socio-Economic Impact Study National Park on Baffin Island, Department of Geography, McGill University.

    MacMillan, J. A., S. Lyon, and N. Brown, Analysis of Socio-Economics Impacts of the Proposed Grasslands National Park, University of Manitoba.

1976   Bhattacharyya, S. K., Regional Economic Analysis of Cape Breton Highlands National Park.

    Foster, Michael, and Andrew S. Harvey, Regional Socio-Economic Impact of a National Park: Before and After Kejimkujik, Institute of Public Affairs, Dalhousie University, Halifax.

    Parks Canada, Socio-Economic Research Division, Socio-Economic Impact Assessment of the Proposed English River System National Park.

1978   Woods Gordon and Co., St. Lawrence Islands National Park Economic Impact Study: Summary Report.

1981   Audet, Michel, L'impact socio-economique du parc national de Kouchibouguac sur son environnement immediat et sur l'ensemble de la Province du Nouveau-Brunswick, prepared for the Special Commission of Inquiry (Federal-Provincial).

    Parks Canada, Socio-Economic Division; Northwest Territories, Economic Development and Tourism, Local Benefits Study: Interim Report.

1982   Parks Canada, Prairie Region and Department of Economic Development and Tourism, Government of Northwest Territories, Economic Impacts of Existing National Parks in the N.W.T.: Summary Report.

National Economic Impact Methodology

1983   Parks Canada Socio-Economic Branch. 1983. *Evaluation of the impact on the Canadian economy of spending attributable to national heritage locations administered by Parks Canada,* vols. I–IV. April 1983.

    Parks Canada, Socio-Economic Branch, Economic Assessment of the Parks Canada Program 1982–83 and 1984–85, December 1985.

    Parks Canada, Prairie Region, Socio-Economic Division, Social and Economic Action Plan: An Examination of the Potential Impacts and Socio-Economic Strategy for the Establishment of a National Park Reserve on Northern Ellesmere.

Capital Project Economic Impact Model

1984   Parks Canada, Socio-Economic Branch, Evaluation of the Economic Impact of Capital Projects Using the Statistics Canada Inter-Provincial Input-Output Model.

    Parks Canada, Socio-Economic Branch, User Guide: Capital Project Economic Impact Model.

    Parcs Canada, Direction socio-economique, Impact socio-economique local et regional de la creation d'un parc national en Minganie: Rapport final.

    Parks Canada, Socio-Economic Branch, User Guide: Socio-Economic Analysis and Impact Assessment in Capital Projects.

Visit Expenditures Model

1984   Beaman, J. The Parks Canada Expenditures Model (see Beaman et al., 1989).

Local Economic Impact Model

1985   Parks Canada, Socio-Economic Branch, User Guide: Tiebout Local Economic Impact Model.

Life-Cycle Cost Model

1985   Parks Canada, Socio-Economic Branch, User Guide: Life Cycle Cost Model.

Woods Gordon, Socio-Economic Impact Assessment of the Proposed National Park on the Bruce Peninsula: Final Report—Text.

Economic Assessment of the Parks Canada Program 1984–85 (update of previous study).

The DPA Group, Proposed West Isles National Marine Park Socio-Economic Impact Study: Final Report.

P. M. Associates Ltd., Manitou Mounds Proposed National Historic Park: Market Analysis and Regional Economic Impact, 1985, prepared for the Socio-Economic Research Unit, Ontario Region, Parks Canada.

Intergroup Consultant Ltd., Economic Performance of Four Northern National Parks/Reserves, prepared for the Evaluation and Analysis Division, Socio-Economic Branch, Parks Canada.

1986   Parks Canada, Socio-Economic Branch, Evaluation and Analysis Division, Economic Assessment of the National Historic Canals 1984–85.

Parks Canada, Socio-Economic Branch, Economic Assessment of St. Peters, Carillon, Chambly, Lachine, Ste-Anne, St-Ours, Rideau, Sault Ste. Marie, Trent-Severn Waterway, Canals 1984–85.

Redekop, D., Regional Economic Impacts of National Parks: Assessing the Dimensions using the LEI Model, Socio-Economic Branch, Parks Canada.

Coriolis Consulting Corporation, "Economic Impact of Fort Langley National Historic Park," Appendix 7 of Fort Langley National Historic Park: Market Analysis and Marketing Strategy, prepared for the Canadian Parks Service, Western Region.

P. G. Whiting and Associates, Economic Models of Environment Canada, Parks, User Requirements Study, prepared for the Socio-Economic Branch, Canadian Parks Service.

P. G. Whiting and Associates, Systems Manual for the Economic Models of Parks, prepared for the Socio-Economic Branch, Canadian Parks Service.

Provincial Economic Impact Model

1987   Soleco Consultants Inc., Provincial Economic Impact Model User Guide, prepared for the Socio-Economic Branch, Canadian Parks Service.

Canadian Parks Service, Socio-Economic Branch, Socio-Economic Analysis for Project Planning: Participants Manual (Module 5: Socio-Economic Impact Assessment and Module 8: Regional Socio-Economic Impact Studies)

Canadian Parks Service, Socio-Economic Branch, Socio-Economic Analysis for Project Planning: Participants Manual.

Rondeau, Daniel, Socio-Economic Branch, Capital Project Economic Impact Model: Review of Applications.

Service canadien des parcs, Section Politiques et recherche (BRQ), Revision du plan de gestion du parc national Forillon, analyse de la frequentation et impact socio-economique des projets de developpementprevus, 1987–1988, 1989–1990.

Butler, James R., et al., The Bird Watchers of Point Pelee National Park, Phase Two: A Socio-Economic Impact Assessment Concerning Birders and the Point Pelee-Leamington District, University of Alberta.

Canadian Parks Service, Ontario Region, Socio-Economic Research Unit, Local Economic Impact Assessment of Pukaskwa National Park.

Canadian Parks Service, Socio-Economic Branch, Using Socio-Economic Branch's Life-Cycle Model.

1988   Strategie Organisation et Methode (SOM), Etude socio-economique sur la creation d'un parc marin national au Saguenay, Volume 3: Evaluation des impacts socio-economiques . . . , prepare pour Environnement Canada, Parcs.

Canadian Parks Service, Socio-Economic Branch, Evaluation and Analysis Division, Socio-Economic Considerations for the Establishment of a National Park on Banks Island: A Preliminary Analysis.

Make or Buy Model

1989   Socio-Economic Branch, Canadian Parks Service, Manual for the review of outside-sector operation and delivery of Parks Services.

Thompson Economic Consulting Service, An Assessment of Induced Investments Resulting from the Purchases of Goods and Services Made by the Canadian Parks Service, Fiscal Year 1988–89, prepared for the Socio-Economic Branch, Canadian Parks Service.

P. G. Whiting and Associates, Private Sector Construction Expenditure Induced by Selected National Heritage Locations in the Atlantic Region, prepared for the Socio-Economic Branch, Canadian Parks Service.

P. G. Whiting and Associates, Economic Impact of Prince Edward Island National Park Non-Resident Visitor Expenditure, prepared for the Socio-Economic Branch, Canadian Parks Service.

P. G. Whiting and Associates, Sault-Ste.-Marie Canal: Socio-Economic Impact Evaluation: Executive Summary, May 1989, prepared for the Socio-Economic Research Unit, Ontario Region, Canadian Parks Service.

Thompson Economic Consulting Services, Visitor Profile and Economic Impact Statement of Northern National Parks (Reserves) and Historic Sites: Summary Report and Technical Appendices, prepared for the Socio-Economic Branch, Canadian Parks Service.

Thompson Economic Consulting Services, Private-Sector Construction in Banff, Jasper and Waterton National Parks in Alberta: Economic Impacts, prepared for the Socio-Economic Branch, Canadian Parks Service.

Canadian Parks Service, Socio-Economic Branch, Impact on the Provincial Economy of the National Parks of Alberta.

1990   Churchill Working Group, Background technical papers related to 1990 report (Appendix 3: Impacts on Tourism and the Economy: Preliminary Analysis, 20 pages).

1992   Gardner Pinfold Consulting Economists Ltd., *An Economic Impact Statement and a Visit Profile of Atlantic Region National Parks and National Historic Sites,* prepared for the Socio-Economic Branch, Canadian Parks Service.

The Coopers & Lybrand Consulting Group, *Economic Impact Analysis of Canals, National Historic Sites and National Parks in Ontario,* prepared for the Socio-Economic Research Unit, Ontario Region, Canadian Parks Service.

Garner Pinfold Consulting Economists Ltd., *Gros Morne National Park Economic Impact Study,* prepared for Parks Canada, 1992.

1993   Stanley, Dick and Bruce Jackson, *The Integrated Business Information System: Using Automation to Monitor Cost Effectiveness of Park Operations,* in Proceedings of the 1994 Northeastern Recreation Research Symposium, April, 10–12, 1994. Saratoga Springs, NY. USDA Forest Service General Technical Report NE–198, pp. 146–151.

Parks Canada (Socio-Economic Branch) Visit Profile and Economic Impact Statement of Manitoba and Saskatchewan National Parks and National Historic Sites: A Summary Report and Technical Appendices, Ottawa.

In June, 1993, Parks Canada was transferred from the Department of the Environment into the newly formed Department of Canadian Heritage. A group of economists from the Socio-Economic Branch left Parks Canada to join the Strategic Research and Analysis Group of the new department. The following outlines the major economic studies undertaken since 1993 as well as recent initiatives to develop a standardized framework for evaluating benefits of protected areas and of activities associated with the arts and cultural sectors.

1994   Strategic Research and Analysis, *A National Park in the North Baffin Area: Potential Impacts on Tourism and the Economics of the Northwest Territories and the Eastern Arctic Region,* prepared for Parks Canada.

Strategic Research and Analysis, *Projected Economic Impacts Associated with the Proposed Torngats National Park in Labrador,* prepared for Parks Canada.

1995   Strategic Research and Analysis, *Guidelines for conducting economic evaluations of major sports events,* Ottawa.

Rogers, Judy (Research Resolutions), *The Economic Impact of the Barnes Exhibit: Final Report and Technical Appendices,* prepared for the Ontario Ministry of Culture, Tourism and Recreation, Toronto.

Whiting, P. G. (The Outspan Group) and Strategic Research and Analysis (Department of Canadian Heritage), *Visit Profile and Economic Impact Statement of Northern National Parks (Reserves) and Historic Sites: A Summary Report,* prepared for Parks Canada, Ottawa.

Standardization of an Economic Valuation Framework of Protected Areas and for Activities Associated with the Arts and Cultural Sectors

1996   Whiting, P. G., Benefits of Protected Areas, prepared for Parks Canada, Ottawa.

Rogers, Judy (Research Resolutions), *Blockbusters and the Cultural Visitor: A Special Analysis of Out-of Town Visitors to the Barnes Exhibit at the Art Gallery of Ontario,* prepared for the Department of Canadian Heritage, Ottawa.

Rogers, Judy (Research Resolutions), *The Northern Tourism Experience: 1994 Domestic and Inbound Markets to Canada's North,* prepared for the Canadian Tourism Commission, the Department of Canadian Heritage, et al.

1997   Stanley, Dick, *Measuring the Benefits of Protected Areas: A Critical Perspective on the IUCN Guidelines,* to be published in the Forthcoming Proceedings of the 1997 Northeastern Recreation Research Symposium, April, 1997, Boulton Landing.

Whiting, P. G. (The Outspan Group), *Benefits of Economic Impacts Associated with the Canadian Heritage Rivers System,* prepared for the Canadian Heritage Rivers Board, Ottawa.

Locke, Wade, *Societal Benefits of Protected Areas: The Gros Morne National Park Case Study,* Prepared for Parks Canada and the Strategic Research and Analysis, Department of Canadian Heritage, Ottawa.

Whiting, P. G. *Projected Economic Impacts Associated with Proposed Wager Bay National,* prepared for Parks Canada.

Statistics Canada, National Accounts and Environment Division, *Tourism Economic Impact Model (TIEM),* prepared for Canadian Tourism Commission (draft, preliminary version), Ottawa.

Rogers, Judy (Research Resolutions), *Renoir's Portraits: Impressions of an Age: Final Report and Technical Appendices,* prepared for the National Gallery of Canada, the Canadian Tourism Commission and the Department of Canadian Heritage, Ottawa.

Provincial Economic Impact Model

1998   The Provincial Economic Impact Model: An enhanced tool for analysts. Spring 1998.

CITED REFERENCES

Acar W. 1976. An integrated approach to multi-dimensional evaluation and cost effectiveness analysis. *CORDS Technical Note No. 25. Canadian Outdoor Recreation Demand Study, vol. II (Technical Notes).* Ottawa: Parks Canada. See also *Public Finance,* 31(1):58–72.

Beaman, J., and S. Grimm. 1989. Park use related data recording: A new direction for the Canadian Parks service. In *Proceedings of the 1989 Northeastern Recreation Research Symposium, April 3–5, 1989, Saratoga Springs, New York.* General Technical Report NE–132, pp. 69–76. Washington, D.C.: USDA Forest Service.

Beaman, J., G. Wall, and M. Cotter. 1989. An introduction to a Visit Expenditure Model for the Canadian Park Service. Ottawa: Parks Canada. The introduction and a 1989 revision of the 1984 monograph is now available only from jaybman@igs.net

Bureau of Management Consulting. 1975 *Tourism Expenditure Model.* Ottawa: Tourism Canada.

Hamilton-Smith, E., and D. Mercer. 1991. Approaches to park planning. In *Urban parks and their visitors,* edited by E. Hamilton-Smith and D. Mercer. Melbourne: Melbourne Water Corporation.

Parks Canada, prepared by P. G. Whiting and Associates. 1987. *Project Options Analysis.* Ottawa: Parks Canada.

Stanley D., J. Beaman, and M. Jaro. 1992. The anatomy of complex data: The underlying structure of questionnaires and forms. In *Proceedings of the 1992 Northeastern Recreation Research Symposium, April 5–7, 1992. Saratoga Springs, New York.* General Technical Report NE–176, pp. 95–98. Washington, D.C.: USDA Forest Service.

Tiebout, C. M. 1962. *The community economic base study.* New York: Committee for Economic Development.

Chapter 12

The Recreation Opportunity Spectrum and the Limits of Acceptable Change Planning Systems: A Review of Experiences and Lessons

George H. Stankey

Growing professionalism in outdoor recreation management over the past twenty years has led to the formulation of more rigorous planning frameworks and approaches. This has been driven by a variety of factors, including the growing body of knowledge in the recreation management field as well as the need to improve the capacity of recreation to compete with other resource uses.

In this chapter, I examine two specific frameworks—the Recreation Opportunity Spectrum (ROS) and the Limits of Acceptable Change (LAC) planning systems. Both arose in response to the conditions described above. Both continue to be used in a variety of settings. The ROS and the LAC are resource management planning systems that have been widely applied in North America and overseas in recent years. This chapter reviews their application, focusing particularly on their use overseas. First, I update the breadth and nature of experience that has been gained in these applications, especially with regard to their use in other countries. A key finding is that overseas experiences have led to a broader range of applications than reported in North America. Second, because the application of both systems continues to be plagued by operational and conceptual difficulties, I close with a discussion of several specific problems that must be addressed in order to achieve successful implementation of either system.

The chapter also identifies some persistent problems in the application of the two systems. These include the failure to identify an underlying rationale, limited integration with other planning systems, a failure to realize the potential of the systems for uses other than inventorying current conditions, and an inadequate basis for indicators and standards.

THE ROS AND LAC: FRAMEWORKS FOR THINKING

The ROS approach derives from the efforts of many people (e.g., Wagar, 1966). However, development of a specific planning approach arose from parallel works undertaken by Driver and Brown (1978) and Clark and Stankey (1979). The concept of a "recreational opportunity setting," defined as the combination of biological, physical, social, and managerial conditions that give value to a place, was integral in both approaches and rested on the premise that recreationists seek a variety of recreational opportunity settings and, through their participation in different activities in these settings, derive a variety of experiences and benefits. By varying the level and character of such setting factors as access, on-site management (e.g., facilities), use density, and amounts and types of regulation, managers could provide a diversity of recreational opportunity settings from which visitors would derive different experiences and benefits.

The works of Driver/Brown and Clark/Stankey also shared a concern with providing managers with a systematic and explicit yet practical approach to making decisions about recreation planning and management. The definition and protection of recreational quality was also a concern, as was the provision of a diversity of recreational settings (Dustin and McAvoy, 1982). Finally, both approaches were intended to provide a systematic framework within which key questions about the provision of recreation settings—what kinds, where, what amounts—could be addressed; both sought to offer a flexible "framework for thinking," rather than a mechanistic, formula-driven approach.

Similarly, the LAC system evolved from an extensive body of management experience and research, deriving from the long-term interest in carrying capacity (Stankey and McCool, 1984). As with the ROS, the LAC is intended to provide managers and planners with a deliberate, conscious, and explicit process to help define the nature and level of changes resulting from recreational use that would be permitted (i.e., accepted) before some type of action was taken. Again, the concept of a framework for thinking is key.

APPLICATIONS OF THE ROS

As a planning framework, the ROS can be implemented in a variety of specific ways. However, the basic process underlying the ROS can be described as follows.

As the name implies, the ROS assumes a continuum of recreation settings, ranging from areas highly developed and intensively used to areas with little evidence of human use and activity and where natural conditions and processes predominate. It is a continuum, as Nash (1982) has aptly described, ranging from "the paved to the primeval." In turn, this focus on a continuum of conditions is based on existing knowledge of the diversity of conditions sought by recreationists as they try to find satisfactory experiences. In other words, diversity in recreational settings is not an end itself, but a means to achieving a range of human benefits.

The focus of a particular application of the ROS could embrace the entire continuum (from the wilderness to a city park) or only a portion thereof. A determination of the appropriate scope of coverage is influenced by such things as an organization's policies, responsibilities, and the nature of resources they administer. Municipal governments, for example, typically would emphasize the developed end of the spectrum; organizations such as the USDA Forest Service give primary attention to the less developed opportunities.

Within the span of the spectrum chosen for emphasis by an organization, a series of classes or "opportunity settings" are selected. As with any classification process, these classes are arbitrary. There is no magic number as to how many classes are "right" or any formula to guide the kinds of conditions they possess or what they are called. The basic intent is to create a series of distinctive settings that are judged to facilitate different kinds of experiences for visitors using them. Thus, for example, if one is concerned with creating opportunities for solitude or challenge, it is likely that access levels or facility development would be minimal. Conversely, if one sought to promote social interaction, improved access and facility development would be appropriate.

By defining specific standards for conditions characterizing each opportunity setting, it becomes possible to map the settings, to compare desired patterns and distributions of opportunity settings against current conditions, and to identify management actions needed to attain or maintain desired conditions.

In summary, the physical output of a ROS analysis is a map that shows different recreation opportunity settings with specifically defined conditions for each setting. Coupled with demand information, such as that found in State Comprehensive Outdoor Recreation Plans (SCORP), managers can make decisions about where different allocations of opportunity classes might be placed. It also facilitates trade-off analysis by allowing one to measure how changes resulting from other resource decisions (e.g., decisions about access and other resource management activities) might affect the desired recreation setting. And it provides managers a better basis for determining how well management programs are meeting the objectives that underlie them. However, the most important output of the ROS likely involves the *thinking process* through which it requires managers and planners to work, the kinds of questions it requires them to consider, and the thoughtful deliberation required for making decisions about the provision of outdoor recreation opportunities.

In the United States, both the USDA Forest Service and USDI Bureau of Land Management have adopted the ROS framework to specific planning procedures for their respective organizations. Likewise, in Canada, the ROS or variants of it have been adopted in both federal and provincial planning. In particular, the Visitor Activity Management Process (VAMP) is related conceptually to the ROS and has been widely adopted at the federal level (Graham et al., 1986).

The ROS has been used in a variety of terrestrial and water-based settings (e.g., Brown et al., 1980; Manning and Ciali, 1981; Stankey and Brown, 1981; Wollmuth et al., 1985). It has also been used to examine interrelationships between recreation and other forest land uses and activities (e.g., Clark and Stankey, 1978; Andreasen, 1982).

More recently, the ROS has gained attention and adoption in a number of other countries, including Australia, the Republic of China, Denmark, New Zealand, and South Africa. In Australia, for example, the ROS has been adopted as a basic framework for recreation planning in settings as diverse as the national parks, multiple-use areas, and public land in both the state of Victoria (Byrne and Vize, 1990) and the Northern Territory (Resources Planning Unit, 1983), and in recreation settings in near-urban areas in the Australian Capital Territory (Richards, 1988). There are also examples of applications of the ROS to riverine and riparian zones involving the integration of recreation as a land use with a range of other objectives, including water supply, flood mitigation, and electricity generation (Pitts and Anderson, 1984; Grieves, 1990). These latter examples are particularly important, given the need to integrate the results of the recreation planning system with other land uses.

As in North America, overseas applications of the ROS continue to be applied predominantly to natural areas. Glavovic (1988), for instance, used it to formulate an integrated land-use planning approach for the administratively designated 70 thousand hectare (173 thousand acre) Cederberg wilderness area north of Cape Town, South Africa. The framework was useful in dealing with the impacts of rapidly increasing numbers of users and as a way to better integrate the area's recreation use with other permitted activities, such as the production of high-quality water.

Overseas applications of the ROS have also helped demonstrate the utility of the framework in settings other than forests, natural parks, and similar areas. For instance, in Australia, Cullen et al. (1987) used the ROS to examine the relationship between the allocation of urban open space to different opportunity settings and the impact of differing allocations upon a fixed operating budget. By identifying minimum funding levels required to maintain the standards for any given opportunity setting, they were able to forecast the effects of changes upon a fixed maintenance budget associated with shifts in the allocation of open space to new opportunity classes. The exercise helped local government officials clarify the objectives regarding both the quantity and quality of open space provision.

Brown (1986) reports on application of the ROS in Denmark in the Jae-gersborg Dyrehave and Hegn urban forests, where three million annual visits are reported on an area of only seventeen hundred hectares (forty-two hundred acres). A major objective was to retain the rural character of the landscape in the face of growing urban pressures. Four opportunity classes were defined to develop a map of the current situation; the resulting map was used to judge the impact of proposed actions within and outside the forest on existing recreation opportunities and to evaluate implementation of management actions that might be taken to respond to changes judged unacceptable.

The Danish example is particularly illuminating in that it demonstrates the adaptability of the ROS framework. As suggested earlier, the predominant expe-rience in applying the ROS in North America has been in natural settings. One consequence of this experience has been the development of a range of specific opportunity classes at the undeveloped end of the spectrum; for example, the U.S. Forest Service identifies four, and sometimes five, categories that typify such settings (primitive, semiprimitive nonmotorized, semiprimitive motorized, roaded natural, and in some regions, roaded modified). The remaining range of settings is embraced by two blanket classifications: rural and urban. However, in the extensively human-modified landscape of Denmark, there is little need for a highly discriminatory system at the natural end of the spectrum, while at the more modified end, there is need for greater discrimination. The Danish example reminds us of the underlying premise of the ROS that a spectrum of environmental conditions exist; moreover, within any one "class" (e.g., rural), a range of diverse conditions can be provided. The particular classes defined by any management organization are simply useful labels; what is appropriate in one place might have little relevance elsewhere. Not every provider has either the capacity or need to provide all settings. Indeed, Denmark possesses very limited ability to offer primitive settings; at the same time, it has the ability as well as the need to look for diversity in the provision of recreation settings in the predominantly rural and urban landscape.

APPLICATIONS OF THE LAC

As noted earlier, interest in the LAC approach to recreation planning has its roots in the long-term concern with defining the carrying capacity of recreation settings. A large body of research, matched by an equally extensive level of management concern, was undertaken, particularly following the end of World War II, to help planners and managers determine the levels at which recreation use should be restricted in order to prevent unacceptable impacts on resources as well as experiences.

However, the search for a measure of a site's carrying capacity proved largely futile. Beginning in the 1980s, much of the research in this area began to refocus its attention away from efforts to measure "how much is too much" to the question of defining what resource conditions (e.g., vegetation, soils, water quality) and social conditions (e.g., level and type of use, location of use) are desired and then undertaking the management actions necessary to ensure those conditions.

The LAC process rests on several fundamental premises. First, it accepts that change in resource conditions is inevitable; natural systems are dynamic and will change irrespective of human actions. Second, the LAC also recognizes that change will result from recreation use; this human-induced change may or may not be appropriate in a given condition, depending upon the specific objectives defined for the area. Third, in managing the change resulting from human use, a variety of actions are possible. Limiting use (which had been the principal focus in carrying capacity studies) is only one action that might be taken to control impacts judged to be unacceptable. Fourth, the judgment of a particular condition as unacceptable is fundamentally a value judgment; hence its determination is a function not only of biological knowledge, but of social choices as well.

As with the ROS system, the LAC represents a conceptual framework within which decisions can be made about appropriate and acceptable levels of impact upon recreation settings and experiences. As such, it is not a mechanistic blueprint, but rather a process that requires conscious, deliberate reflection. It rests on four major components: (1) the specification of acceptable and achievable resource conditions, as defined by a measurable set of indicators (e.g., level of vegetation loss at campsites, level of encounters while traveling along trails); (2) an analysis of the relationship between existing conditions and those judged acceptable (simply put, a comparison between what is desired and what currently exists); (3) identification of management actions necessary to ensure that desired conditions are achieved or maintained; and (4) a program of monitoring and evaluating management effectiveness.

The LAC planning system has received wide attention and growing application. In the United States, both the Forest Service and Bureau of Land Management use the system, particularly in wilderness planning. However, although the original document (Stankey et al., 1985) specifically described the LAC as a ''wilderness planning system,'' in actuality it is a generic planning framework, with potential applications across a wide range of issues, including the management of such nonrecreation resource issues as air and water quality. In North America, however, most applications have focused on recreation in dispersed, undeveloped settings.

In general, this is also the case overseas. As compared with the ROS, the LAC has not been as widely used in part because it is newer. However, examples of application can be found in Australia, Britain, New Zealand, Norway, the Republic of China, and South Africa. In Australia the LAC has been used in a manner similar to that in the United States. It has been proposed as a fundamental framework to guide planning and management of the Australian Alps National Park, an area managed under a tripartite agreement among the states of New

South Wales, Victoria, and the Australian Capital Territory (Cullen and Turner, 1986). It has also been used in development of a management strategy for the 94 thousand hectare (232 thousand acre) Namadgi Nature Reserve near Canberra, Australia. The reserve serves as a water catchment for Canberra; nondegradation of water supply and quality are key issues, and the question of "acceptable" impact upon water quality resulting from recreational use within the Reserve has been a concern (Turner, 1988).

The Australians also have provided examples of the LAC in nonwilderness as well as nonrecreational settings. Cullen (1988) used the LAC as a framework for assessing the impacts of discharging secondary sewage to a mountain stream; his work reveals the key importance of selecting appropriate indicators of environmental change and the need to establish both the season and annual variability of those indicators. While in Australia between 1987 and 1989, the author became involved in a project examining the association between vegetation removal in arid regions in western New South Wales and associated impacts on salt infiltration into water tables; the LAC provided the basic framework for this analysis.

PROBLEMS AND SHORTCOMINGS IN APPLICATION OF THE ROS AND LAC

This brief review of applications of the ROS and LAC suggests they have proved useful in a variety of functional and institutional settings. Moreover, they appear to have utility when applied in non–North American situations. At the same time, it would be incorrect to leave the impression that problems and difficulties regarding the use of these systems have been overcome. Indeed, many problems persist (Clark, 1982); as we shall see in this final discussion, many of these problems derive from the failure to apply the systems consistently with the premises and assumptions that underlie them, making them vulnerable to criticism (e.g., van Oosterzee, 1984).

A LACK OF RATIONALE

The most persistent, and in many ways, most serious shortcomings in applications of either the ROS or LAC is the failure to identify an explicit rationale. A rationale constitutes an underlying reason and purpose for an activity, a set of controlling principles upon which practices are based. Too often, however, efforts to apply these systems turn immediately to developing indicators and standards, drawing lines on maps, or other specific procedures before clearly identifying these underlying reasons and purposes in any explicit fashion. But without a clearly specified rationale, there exists no basis upon which the judgments required to implement or evaluate the systems can be based. This is particularly serious, given that both the ROS and LAC are "state of knowledge" planning systems; i.e., they rely upon judgments based on our best understanding of current knowledge, using these judgments as a basis for proposed actions while allowing for revision as new and better information becomes available. However, when the decision rationale is unclear, ambivalent, or lacking altogether, such judgments have little apparent basis and might appear as little more than personal preference or whim.

For example, many ROS plans identify "remoteness" as one criterion for specifying opportunity classes. As a standard for defining remoteness, usually some measure of the distance from roads, railroads, or other permanent development is involved; the U.S. Forest Service guidelines, for instance, identify an area as a "primitive opportunity class" if it is at least three miles from such developments. But why three miles? Why not one, or five, or ten? Without an underlying rationale for what one is attempting to accomplish, it is impossible to justify (or refute) any measure. However, if one accepts the rationale that the criterion is intended to help buffer the individual from the sights and sounds of civilization and that, all other things being equal, in the generally forested and mountainous regions that characterize many national forests, a three-mile buffer provides this, the standard becomes intelligible and defensible. Given a similar rationale but applied to a desert or forested coastal zone, an appropriate standard might be as much as fifteen miles or as little as a few hundred yards.

The failure to specify a rationale for either the ROS or LAC also has contributed to a tendency to "borrow" the classes, indicators, and even standards used in other applications. At best, this can lead to inappropriate or inapplicable conditions; at worst, it can lead to fatal flaws in the application of either system. While it is entirely appropriate to review other applications as a source of understanding and learning about how the processes were implemented, the wholesale, undiscriminating adoption of specific features of other applications can lead to serious problems and ultimately will make implementation impossible.

The failure to define a rationale leads to many associated problems. It tends to confuse means with ends. It fosters a mechanistic approach to both the ROS and LAC, seeking simplicity and universality rather than adapting to site- or area-specific conditions. It tends to focus attention on finding answers rather than solutions to problems. And it reflects a "cookbook" approach to planning, seeking formulas and, as Socolow (1976) has described, "golden rules and golden numbers" as a substitute for critical thinking. Conversely, a sound rationale helps planners ask the right questions and provides them with a reasoned, defensible basis for the judgments that the ROS and LAC require if they are to be effective.

INTEGRATION WITH OTHER PLANNING SYSTEMS

The ROS is a framework for thinking about recreation management; the LAC has wider potential, but most applications have likewise focused on recreation. Neither system is intended to be, or capable of being, a total resource planning framework (Clark, 1982). Both should, however, represent a means of integrating recreation values with other resource management considerations.

Unfortunately, there have been few efforts to establish such integrative linkages with other resources or planning systems. Most planning remains functional and reductionist. These limitations must be addressed, in part, because only then will we be able to ensure that adequate and equitable attention is given to recreation management and that opportunities for truly integrating recreation in resource management can be realized. The focus of the ROS and LAC on settings and specific resource conditions provides the basis for a common language that resource managers can use to discuss relationships between different activities (Clark, 1988).

There is some evidence of movement in this direction. For instance, the Forest Service's Visual Management System (VMS) has recently been revised, with particular attention given to ways in which it can be better integrated with the ROS system. The LAC also has been linked to visual management considerations by Bureau of Land Management managers to create the Visual Impact Evaluation System (Overbaugh, 1990). The LAC also has been examined in terms of how it can better integrate fish and wildlife management concerns (Montana Department of Fish, Wildlife, and Parks, 1988). Clark (1988) has described ways in which silviculture can be used to enhance recreation opportunities, using many of the concepts and terms derived from wildlife management.

WHAT IS VERSUS WHAT SHOULD BE

A recurring problem in applications of the ROS is that current conditions come to dominate the question of future possibilities. A ROS inventory provides a baseline of existing conditions; it describes "what is," a necessary and fundamental part of any planning process. However, this is only the first phase of the ROS process; other phases focus on "what can be," "what should be," and "what will be" (Stankey et al., 1983). The present condition does not dictate the choices to be considered in the other phases. It obviously has influences, but the choice always remains to alter "what is" to something else. The capacity to do this will depend upon a variety of factors, including budget and knowledge.

Part of the reason that this problem persists is that the ROS has remained largely a supply-driven concept. We usually have considerable ability to define recreation opportunities in supply terms, but our ability to examine the demand side of the equation is much less well developed. Knowledge about public preferences, values, and needs is often sketchy or absent altogether; if data are available, they typically are expressed in activity terms (e.g., camping, fishing), which have limited value in the ROS. The lack of appropriate demand data makes efforts to prescribe "what should be" more difficult. Often the presumption that what we have represents what is desired dominates, even though there is little basis for such an assumption. There is a major need to improve our measures of public values, preferences, and requirements about recreation in order to fully capitalize on what the ROS framework can provide.

BASIS OF INDICATORS AND STANDARDS

Both the ROS and LAC systems use indicators and standards as a means of operationalizing and measuring desired conditions and as the "trigger" to define when certain management actions are required. Information to define indicators and set standards is often sketchy, and professional judgments, as noted earlier, become very important, lest the system become paralyzed. However, a number of problems can be noted.

First, there is no universal list of indicators, although an effort to develop a comprehensive inventory of them has been made (Merigliano and Krumpe, n.d.). However, the LAC in particular is an issue-driven process; that is, the indicators that are selected to implement the LAC derive from an analysis of the issues and concerns identified by interested citizens and managers during the planning process itself. What this means in practice is that different indicators will emerge, often only relevant in a particular situation. However, because of the tendency to select indicators that have been used elsewhere, there is the likelihood that the resulting planning effort will focus on issues that might or might not be of consequence in the particular situation under consideration.

Second, because standards are often based on limited data and understanding, uncertainty inevitably surrounds them. For example, if a standard is established for the level of water pollution that can be tolerated before human health is endangered, how confident can we be that, in fact, that level will not be injurious to our health? Systematic monitoring and evaluation (ME) are essential sources of feedback to managers as to the soundness of their judgments and the need for changes, either in the standards (in the event of being founded upon inadequate understanding) or in the management practices that have been undertaken. Too often, however, ME occurs only spasmodically or not at all. The lack of rigorous ME jeopardizes effective application of both the LAC and ROS systems.

Third, because the rationale for these planning systems is often vague or unstated, the question of what appropriate indicators and standards are is itself difficult to answer. For example, concern with water quality in backcountry areas has often been addressed by monitoring for the presence of human health pathogens. The general failure to find such organisms led many to assume there was no serious impact on water quality as a result of recreation use in these areas. However, when attention focused on water chemistry, it was discovered that significant problems had developed associated with the import of nutrients by recreationists (Silverman and Erman, 1970). Focusing on an inappropriate indicator and its condition (human health) resulted in the failure to identify a significant impact upon one of the major values (i.e., naturalness) for the area.

Fourth, because the basis for standards is often not well supported and the assumptions underlying them are poorly defined or absent altogether, there can be a temptation to respond to growing use pressures by changing the standard, as opposed to implementing new, perhaps more restrictive, management actions. Although standards should not be viewed as sacrosanct (a fundamental characteristic of standards is that they should be achievable), neither should they be routinely altered whenever conditions approach them. Indeed, their major function is to alert managers to the need for action in order to make possible the continued provision of a particular set of conditions defined through a planning process as important. To continually alter them as a means of avoiding the implementation of what might be seen as more difficult or controversial management programs compromises the core of both the ROS and LAC.

THE LINKAGE BETWEEN SETTINGS AND EXPERIENCES

A more fundamental conceptual area of debate focuses on the question of the linkage between specific ROS settings and their capacity to produce specific desired experience outcomes. A characteristic of the ROS framework, particularly as conceptualized by Driver and Brown (1978), was its similarity to a production system; the input of investments, knowledge, visitor use, and so forth lead to the production of specific outcomes, in this case, experiences. Certainly, at a general level of resolution, this is the case. However, debate continues about the predictability of this relationship and the underlying issue of the causality involved.

First, the relationship between visitor use of particular settings and the production of particular experiences from those settings has always been seen to be of a probabilistic nature. Second, certain experiences are more likely to be associated with the nature of settings than others; as Driver et al. (1987) have noted, an assumption of the ROS (and one requiring additional empirical support) was that the realization of some experiences, such as isolation, large-group affiliation, and risk, depended upon the availability of particular combinations of activities and settings.

This assumption has gained some support. For example, research in a variety of recreation settings confirms that the setting-experience linkage is indeed highly situationally specific, dependent upon the types of activities involved, the specific experiences involved, the level of prior experience of the visitor, and so on (Virden and Knopf, 1989; Yuan and McEwen, 1989; Heywood, 1991; Heywood et al., 1991). One conclusion that emerges from this research is a confirmation of the complex, nonlinear, and multifaceted nature of the relationship between setting and experiences, a feature supporting Manning's (1985) concern that rigid, inflexible applications of the ROS might reduce, rather than enhance, diversity in recreational settings. To the extent that ROS applications come to be driven by a mechanistic, formula-based mentality, this will remain a legitimate concern.

Finally, the question of the linkage between settings and experiences is confounded by the question as to what specific features of the setting are, in fact, being measured. The ROS system has relied upon objective inventories of specific resource conditions or attributes in order to be operationalized. Whether these specific conditions are salient to the choice process of recreationists, however, is another matter. Clark and Stankey (1978) note that one criterion to guide the selection of management factors in the ROS is their relationship to the visitor's choice process; unless the factor is important or salient to visitors, variations in its condition from one opportunity setting to another will have little or no significance to the visitor.

The relationship between settings and experiences clearly requires further study. Three aspects warrant specific mention here. First, the perception of what attributes of recreation settings are important clearly differs between visitors and managers as well as among visitors (e.g., Heywood, 1991). Greater attention to the definition of attributes related to the visitor choice process is needed (Clark and Stankey, 1985).

Second, the constructs we use to measure the meaning and complexity of visitor experiences (e.g., solitude, challenge) are crude and unsophisticated. For example, Virden and Knopf (1989) found that the opportunity to "experience nature" was largely independent of either activity or setting preference. However, we lack the ability to determine exactly what an expressed interest in an outcome such as "experiencing nature" really involves. Similarly, experience outcomes such as solitude, risk, and socialization appear important across a range of recreation opportunity settings, but we understand little about how the specific qualities of these constructs might vary within settings or how management activities might either facilitate or thwart their realization.

Third, it is important that we not lose sight of the fact that experiences remain the output of recreation management. Despite the difficulties in developing specific and predictable measures of how changes in settings might (or might not) affect visitor experiences, managers and researchers need to continue to pay attention to ways in which diverse recreation settings can be provided and to creative ways in which diversity can be most appropriately defined.

ACCEPTABILITY: DESIRED GOAL OR MINIMUM TARGET?

The Limits of Acceptable Change planning system was founded on the premise that change is a natural and inevitable characteristic of settings. However, human-induced change, resulting from recreation, can lead to alterations in natural systems that might not be desirable from the point of view of area management objectives. In wildernesses, for example, the level and extent of human-induced change can conflict with objectives related to the protection of ecological processes and functions.

Although the original paper outlining the LAC concept (Stankey et al., 1985) did not explicitly define acceptability, the report suggests two differing conceptions. First, it is used to describe what is legally permissible under the Wilderness

Act of 1964. Second, it is used to describe what wilderness users agree is desirable as expressed through a consensus-based planning approach.

Thus acceptability can be viewed as either the minimally acceptable conditions that can be permitted or it can be used as the statement of a desirable state for which management strives. These are very different conceptions, both operationally as well as philosophically. In the former sense, the focus is basically on what level of change can we accept before some type of severe consequence reveals itself, be it ecological (e.g., pollution of a stream leading to a loss of fish life) or social (e.g., conditions reach a point that users take legal action). In the latter sense, acceptability describes a desired goal: a set of conditions judged appropriate (i.e., acceptable).

Although the concept of acceptable change has usefully refocused attention in recreation management from "how much is too much" to "what are the conditions we wish to maintain or restore," much remains problematic about the idea. Difficult questions remain, related to such things as who makes such decisions, the relevant variables for which decisions about change are made, as well as questions about the trade-offs involved in accepting one form of change as opposed to another (e.g., change in ecological conditions versus change in social conditions).

When acceptability is used as a means of reducing dissent to tolerable levels while adhering to the status quo as closely as possible, public cynicism can be anticipated (Brunson, 1993). Its use as a means of "just getting by" will largely negate the potential value it has contributed to questions about the management of recreation impact in either ecological or social terms.

CONCLUSIONS

The management of outdoor recreation resources remains a major challenge. Long treated as a residual activity, requiring little other than a custodial level of management attention, its growing importance as a social and economic activity requires a significant increase in the level of investment, sophistication, and attention it receives.

The ROS and LAC planning systems are examples of responses to the need to enhance the quality and level of recreation management. With companion systems such as VAMP and the Visitor Impact Management System (Graefe et al., 1990), recreation managers have available to them an increasingly useful set of tools with which to deal with the major challenges confronting them.

But such tools are only as valuable as the level of commitment and energy given to them. None are substitutes for thinking, for reasoned judgments, or for thoughtful discourse among recreation managers and their research colleagues as well as the broader community of interests. Indeed, their success is wholly dependent upon the exercise of these qualities.

REFERENCES

Andreasen, B. 1982. Utilizing the recreation opportunity spectrum as a tool in land allocation decisions involving energy development and wildlife. In *Issues and technology in the management of impacted western wildlife,* 43–46. Steamboat Springs, Colorado: Thorne Ecological Institute.

Brown, P. J. 1986. About the recreation opportunity spectrum and three reports on the application of that technique for recreation planning and management in Denmark. Corvallis: College of Forestry, Oregon State University.

Brown, P. J., B. L. Driver, and J. K. Berry. 1980. Use of the recreation opportunity planning system to inventory recreation opportunities of arid lands. In *Arid land resource inventories: Developing cost-efficient methods,* 123–128. General Technical Report WO–28. Washington, D.C.: USDA Forest Service.

Brunson, M. W. 1993. "Socially acceptable" forestry: What does it imply for ecosystem management? *Western Journal of Applied Forestry* 8(4): 116–119.

Byrne, N., and R. Vize. 1990. An inventory of recreation opportunity settings on major areas of public land in Victoria. Occasional Paper Series NPW No. 4, 58 pp. Melbourne, Victoria: National Parks and Wildlife Division, Department of Conservation and Environment.

Clark, R. N. 1982. Promises and pitfalls of the ROS in resource management. *Australian Parks and Recreation* (May): 9–13.

Clark, R. N. 1988. Enhancing recreation opportunities in silvicultural planning. In *Proceedings—future forests of the Mountain West: A stand culture symposium,* W. C. Schmidt, compiler, 61–69. Ogden, Utah: USDA Forest Service, Intermountain Research Station.

Clark, R. N., and G. H. Stankey. 1978. Determining the acceptability of recreational impacts: An application of the outdoor recreation opportunity spectrum. In *Conference proceedings: Recreational impact on wildlands,* edited by R. Ittner, D. R. Potter, J. K. Agee, and S. Anschell, 32–42. Portland, Oregon: USDA Forest Service, Region 6.

Clark, R. N., and G. H. Stankey. 1979. The recreation opportunity spectrum: A framework for planning, management, and research. General Technical Report PNW–98. Portland, Oregon: USDA Forest Service, Pacific Northwest Forest and Range Experiment Station.

Clark, R. N., and G. H. Stankey. 1985. Site attributes—A key to managing wilderness and dispersed recreation. In *Proceedings—National Wilderness Research Conference: Current Research,* Robert C. Lucas, compiler, 509–515. USDA Forest Service General Technical Report INT–212. Ogden, Utah: USDA Forest Service, Intermountain Research Station.

Cullen, P. 1988. Predicting the magnitude and acceptability of environmental change. Paper presented to the VII Annual Conference of the International Association for Impact Assessment, Brisbane, Australia.

Cullen, P., I. Oelrichs, G. O'Dell, and K. Yates. 1987. Planning and management of open space. Paper presented at the 60th National Conference, Royal Australian Institute of Parks and Recreation, Canberra, Australia.

Cullen, P., and A. Turner. 1986. Planning for recreation in the Australian Alps National Park. Discussion paper prepared for the Recreation Planning Workshop, Canberra College of Advanced Education, Australia.

Driver, B. L., and P. J. Brown. 1978. The opportunity spectrum concept and behavior information in outdoor recreation resource supply inventories: A rationale. In *Integrated inventories of renewable natural resources: Proceedings of the workshop,* 24–31. Fort Collins, Colorado: USDA Forest Service, Rocky Mountain Forest and Range Experiment Station.

Driver, B. L., P. J. Brown, G. H. Stankey, and T. G. Gregoire. 1987. The ROS planning system: Evolution, basic concepts, and research needed. *Leisure Sciences* 9:201–212.

Dustin, D. L., and L. H. McAvoy. 1982. The decline and fall of quality recreation opportunities and environments? *Environmental Ethics* 4:49–57.

Glavovic, B. C. 1988. A proposed framework for recreation planning in South Africa, with particular reference to the Cederberg. Thesis for Master of Science in Environmental and Geographical Science, 178 pp. University of Cape Town, Cape Town, South Africa.

Graefe, A. R., F. R. Kuss, and J. J. Vaske. 1990. Visitor impact management: The planning framework, vol. II, 105 pp. Washington, D.C.: National Parks and Conservation Association.

Graham, R., P. Nilsen, and R. J. Payne. 1986. A marketing orientation to visitor activity planning and management. Paper presented to the First National Symposium on Social Science in Resource Management, Oregon State University, Corvallis.

Grieves, J. 1990. The impact of recreation on public land—with special reference to river frontage. Master of Environmental Studies, 69 pp. School of Environmental Planning, University of Melbourne, Victoria.

Heywood, J. L. 1991. Visitor inputs to recreation opportunity spectrum allocation and management. *Journal of Park and Recreation Administration* 9(4): 18–30.

Heywood, J. L., J. E. Christensen, and G. H. Stankey. 1991. The relationship between biophysical and social setting factors in the recreation opportunity spectrum. *Leisure Sciences* 13:239–246.

Manning, R. E. 1985. Diversity in a democracy: Expanding the recreation opportunity spectrum. *Leisure Sciences* 7(4): 377–399.

Manning, R. E., and C. P. Ciali. 1981. Recreation and river type: Social-environmental relationships. *Environmental Management* 5(2): 109–120.

Merigliano, L. L., and E. Krumpe. n.d. The identification and evaluation of indicators to monitor wilderness conditions, 33 pp. Moscow: University of Idaho, Forest, Wildlife, and Range Experiment Station.

Montana Department of Fish, Wildlife, and Parks. 1988. Fish and wildlife of the Bob Marshall Complex and surrounding area: Limits of acceptable change in wilderness, 161 pp. Helena, Montana: Montana Department of Fish, Wildlife, and Parks.

Nash, R. 1982. *Wilderness and the American mind,* 3rd ed., 425 pp. New Haven, Connecticut: Yale University Press.

Overbaugh, W. L. 1990. Visual impact evaluation system: A visual limit of acceptable change. In *Managing America's enduring wilderness resource,* edited by D. Lime, 210–213. St. Paul: University of Minnesota.

Pitts, D. J., and D. R. Anderson. 1984. Lake Wivenhoe: An example of comprehensive water storage planning. Paper prepared for Australian Water Resources Council Workshop.

Resources Planning Unit. 1983. A proposal for the adoption of a Commissionwide "Recreation resources and visitor information system," 14 pp. Darwin: Conservation Commission of the Northern Territory.

Richards, G. P. 1988. Outdoor recreation: The opportunity spectrum approach applied in the Australian Capital Territory. Thesis for Master of Applied Science, 270 pp. Canberra College of Advanced Education, Canberra, Australia.

Silverman, G., and D. C. Erman. 1979. Alpine lakes in Kings Canyon National Park, California: Baseline conditions and possible effects of visitor use. *Journal of Environmental Management* 8:73–87.

Socolow, R. H. 1976. Failures of discourse: Obstacles to the integration of environmental values into natural resource policy. In *When values conflict: Essays on environmental analysis, discourse, and decision.* edited by L. H. Tribe, C.S. Schelling, and J. Voss, 1–33. Cambridge, Massachusetts: Ballinger Publishing.

Stankey, G. H., and P. J. Brown. 1981. A technique for recreation planning and management in tomorrow's forests. In Proceedings of Division 6, XVII IUFRO World Congress, 63–73. Tokyo: International Union of Forest Research Organizations.

Stankey, G. H., P. J. Brown, and R. N. Clark. 1983. Monitoring and evaluating changes and trends in recreation opportunity supply. In *Renewable resource inventories for monitoring changes and trends: Proceedings of an international conference,* edited by J. F. Bell and T. Atterbury, 227–230. Corvallis: Oregon State University, College of Forestry.

Stankey, G. H., D. N. Cole, R. C. Lucas, M. E. Peterson, and S. S. Frissell. 1985. The limits of acceptable change (LAC) system for wilderness planning. General Technical Report INP-176, 37 pp. Ogden, Utah: USDA Forest Service,Intermountain Forest and Range Experiment Station.

Stankey, G. H., and S. F. McCool. 1984. Carrying capacity in recreational settings: Evolution, appraisal, and application. *Leisure Sciences* 6(4): 453–474.

Turner, A. 1988. Impact, damage, and management: The place of measurement in public policy making. Paper presented at the Institute of British Geographer's Annual Conference, Loughborough University of Technology, England.

van Oosterzee, P. 1984. The recreation opportunity spectrum: Its use and misuse. *Australian Geographer* 16(2): 97–104.

Virden, R. J., and R. C. Knopf. 1989. Activities, experiences, and environmental settings: A case study of recreation opportunity relationships. *Leisure Sciences* 11:159–176.

Wagar, J. A. 1966. Quality in outdoor recreation. *Trends* 3(3): 9–12.

Wollmuth, D. C., J. H. Schomaker, and L. C. Merriam Jr. 1985. River recreation experience opportunities in two recreation opportunity spectrum (ROS) classes. *Water Resources Bulletin* 21(5): 851–857.

Yuan, M. S., and D. McEwen. 1989. Test for campers' experience preference differences among three ROS setting classes. *Leisure Sciences* 11:177–185.

INDEX